Seed Dormancy and Germination

D0061195

TERTIARY LEVEL BIOLOGY

A series covering selected areas of biology at advanced undergraduate level. While designed specifically for course options at this level within Universities and Polytechnics, the series will be of great value to specialists and research workers in other fields who require a knowledge of the essentials of a subject.

Recent titles in the series:

Biology of Reptiles	Spellerberg
Biology of Fishes	Bone and Marshall
Mammal Ecology	Delany
Virology of Flowering Plants	Stevens
Evolutionary Principles	Calow
Saltmarsh Ecology	Long and Mason
Tropical Rain Forest Ecology	Mabberley
Avian Ecology	Perrins and Birkhead
The Lichen-Forming Fungi	Hawksworth and Hill
Plant Molecular Biology	Grierson and Covey
Social Behaviour in Mammals	Poole
Physiological Strategies in Avian Biology	Philips, Butler and Sharp
An Introduction to Coastal Ecology	Boaden and Seed
Microbial Energetics	Dawes
Molecule, Nerve and Embryo	Ribchester
Nitrogen Fixation in Plants	Dixon and Wheeler
Genetics of Microbes (2nd edn.)	Bainbridge
Seabird Ecology	Furness and Monaghan
The Biochemistry of Energy Utilization in Plants	Dennis
The Behavioural Ecology of Ants	Sudd and Franks
Anaerobic Bacteria	Holland, Knapp and Shoesmith
An Introduction to Marine Science (2nd edn.)	Meadows and Campbell

TERTIARY LEVEL BIOLOGY

Seed Dormancy and Germination

J.W. BRADBEER, BSc, PhD, DSc
Professor of Botany
King's College London

Blackie

Glasgow and London

Published in the USA by
Chapman and Hall
New York

Blackie and Son Limited,
Bishopbriggs, Glasgow G64 2NZ
7 Leicester Place, London WC2H 7BP

Published in the USA by
Chapman and Hall
in association with Routledge, Chapman and Hall
29 West 35th Street, New York, NY 10001—2291

© 1988 Blackie and Son Ltd
First published 1988

All rights reserved
No part of this publication may be reproduced, stored in a retrieval system, or transmitted, in any form or by any means, electronic, mechanical, recording or otherwise, without prior permission of the Publishers.

British Library Cataloguing in Publication Data

Bradbeer J.W.
Seed dormancy and germination
1. Seeds. Dormancy and germination
I. Title II. Series
582′.0333

ISBN 0-216-91635-6
ISBN 0-216-91636-4 Pbk

Library of Congress Cataloging-in-Publication Data

Bradbeer, J.W.
Seed dormancy and germination.

(Tertiary level biology)
Bibliography: p.
Includes index.
1. Germination. 2. Seeds—Dormancy. 3. Seeds.
I. Title. II. Series.
QK740.B73 1988 582′.0333 88-4296
ISBN 0-412-00611-1
ISBN 0-412-00621-9 (pbk.)

Photosetting by Thomson Press (I) Ltd.
Printed in Great Britain by Bell & Bain (Glasgow) Ltd.

Preface

The germination of seeds is a magical event, in which a pinch of dust-like material may give rise to all the power and the beauty of the growing plant. The mechanisms of seed dormancy, of the breaking of seed dormancy and of germination itself continue to remain shrouded in mystery, despite the best efforts of plant scientists. Perhaps we are getting there, but very slowly.

This book considers germination and dormancy from the point of view of plant physiology. Plant physiologists attempt to understand the relationship between plant form and function and to explain, in physical and chemical terms, plant growth and development. The place of germination and dormancy in plant ecophysiology is taken into account with attempts to understand the seed in its environment, whether the environment be natural, semi-natural or wholly artificial. In due course plant scientists hope to develop a precise understanding of germination and dormancy in cellular and molecular terms, and therefore there is some biochemistry in this book. Biochemists who wish to learn something about seeds should find this book useful.

As one who claims to be both a botanist and a biochemist, I am convinced that any understanding of seed function requires a knowledge of seed structure. A consideration of structure and of structural aspects of seed formation and germination find an early place in the book. However, the experimental approach is stressed wherever it is appropriate, and the book concludes with suggestions for the experimental investigation of seed ecophysiology. As this is a textbook, not a review, I have taken many of my examples from my own laboratory, and have selected the remainder from the most significant in the literature.

This book was turned from possibility into reality through the unfailing help and support of Dr E.M. McMorrow who ran my research and teaching programme on seeds for several seasons, who organized and conducted our joint survey of the literature and helped greatly with discussion and criticism. I owe much gratitude to Professor P.F. Wareing FRS who encouraged me to become involved in research on seed dormancy and whose enthusiasm was highly infectious. In the course of my research I have relied on a substantial number of technicians, postgraduate students and postdoctoral fellows whose efforts I gratefully acknowledge. In order of arrival in my laboratory they are: Dr B. Colman, Dr N.J. Pinfield,

Dr A. Wood, Miss V.M. Floyd, Dr B.C. Jarvis, Mr K. Maybury, Dr J.D. Ross, Dr S. d'Apollonia, Dr K.A. Bhagwat, Mrs J. Stubbs, Dr P.M. Williams, Dr Ingrid Arias, Mrs L. Langman, Mrs Nirmala Rajkumar, Mr S.A. Hamid, Dr A.L. Khandakar, Dr C.A. Anon, Dr G.A. Rendon, Mrs B.N. Rogers-Halliday and Mr D.C. James. Amongst the many seed scientists in other laboratories who have provided help and support I must mention my former colleague Dr D.R. Murray of the University of Wollongong, New South Wales.

Finally, I wish to acknowledge the consistent and unfailing support of the publishers.

JWB

To Mary

Contents

CHAPTER ONE
THE SEED

The seed is the product of the fertilized ovule, which in the gymnosperms is naked', being borne on the surface of the scales that comprise the cone, whereas in the angiosperms the seed is formed within an ovary. In time the seed will normally become separated from the parent plant and eventually may germinate to give a new individual.

The role of the seed

In seed production, plants may be regarded as achieving up to four main objectives, namely, a re-sorting of their genetic material, a dispersal mechanism, a multiplication mechanism and a survival mechanism.

In the first case, seed production normally results in an opportunity for the segregation and recombination of the genes so that the seed population contains some new genetic combinations. Of course, in outbreeding populations a greater range of new combinations tends to occur than in inbreeding populations. Furthermore, where seed production is the result of one of the apomictic processes which excludes meiotic segregation, then any new genetic combinations in the seed population can only be produced by mutation. For further information, texts on basic genetics, population genetics and population biology should be consulted. In population biology, the texts by Harper (1977) and Grime (1979) are particularly relevant to many aspects of the present work.

The release of seeds from the parent plant normally results in their dispersal over a comparatively small radius, although, as discussed in Chapter 7, wide dispersal is possible. Wide dispersal of a large number of seeds provides the species either with the opportunity to multiply the size of its population, or, if niches suitable for colonization are very infrequent, with the opportunity of perhaps maintaining its population.

The main property of seeds, that of being able to survive adverse environmental conditions, often for many years, provides the main theme for the present book. Consideration will be given to seed survival under natural conditions (Chapter 7), to the morphological and physiological mechanisms which permit such survival (Chapters 3, 4, 5 and 6) and to the utilization of these mechanisms by man, in what often turn out to be contrasting roles as agriculturalist and as conservationist of the environment and of the seed plants. It seems appropriate to comment on the ability of most seeds to withstand environmental conditions far beyond the tolerance of the whole plant. The resistance of seeds to environmental extremes increases as seeds lose water during and after the final stages of development on the parent plant. However, all seeds have a minimum moisture content below which irreversible damage occurs. Seeds with hard impermeable coats lose water more slowly than seeds with soft coats, and consequently hard seeds tend to be most resistant to desiccation and extremes of temperature. Table 1.1 gives an indication of the considerable resistance of seeds to high-temperature treatments. Crop-plant seeds which tend to have comparatively soft coats take several hours to lose 50% of their viability as a result of exposure, in the dry state, to temperatures of 70 °C or 80 °C. Beadle (1940) showed that all of the seeds that he found and studied in an Australian forested area, which was quite frequently subjected to bush fires, were able to survive several hours of dry heat at 100 °C, with some seeds being able to withstand a period at 120 °C or 130 °C. He also showed that dry seeds are resistant to hot water, although the survival times were much shorter than for dry heat (Table 1.1). As bush fires are a common natural event (as well as a man-made one), these laboratory experiments demonstrate the known phenomenon of seed survival in the upper layers of the soil, which permits seed germination and seedling survival after a fire. Some seeds require to experience a fire before

Table 1.1 The effects of short exposures to high temperatures on the viability of air-dry seeds of a number of species (after Beadle, 1940).

Species	Temperature (°C) at which a 50% loss of seed viability occurs	
	Dry heat for 4 h	Immersed in hot water for 5 min
Pea	70–80	70–80
Wheat	90–100	50–60
Sunflower	60–70	60–70
Hakea acicularis	110–120	70–80
Casuarina rigida	110–120	70–80
Leptospermum scoparium	120–130	70–80
Banksia serrata	–	70–80
Acacia decurrens[1]	100–110	> 100

[1] Hard seed coat

germination will occur (e.g. Warcup, 1980) while others (such as *Banksia ericifolia*) do not dehisce the seeds from persistent fruits until the parent plant has been subjected to fire (Bradstock and Myerscough, 1981).

Air-dry seeds normally show great resistance to very low temperatures, being able to survive long exposures to freezer temperatures (− 20 °C) and liquid N_2 (− 196 °C) (see Chapter 8). Even imbibed seeds in wet soil may show considerable resistance to freezing temperatures. In lettuce, the mechanism of this resistance appears to lie in the ability of the embryo to undergo supercooling without the formation of ice crystals, with the endosperm acting as a barrier to the internal propagation of ice crystals from external nucleation (Keefe and Moore, 1981, Bourque and Wallner, 1982).

Germination

The eventual function of the surviving seed is its germination, followed by the growth of the embryo to give a mature plant. Although an air-dry seed may show a barely perceptible metabolism, the metabolism of an imbibed seed can normally be clearly demonstrated. The germination process is presumed to commence with sequences of events at the molecular and cellular levels which precede visible growth of the embryo. It is usual to record the protrusion of the radicle through the seed coat as the measurable occurrence of germination, although there are certain circumstances where another criterion, such as the geotropic curvature of the radicle of an isolated embryo, has had to be used (Webb and Wareing, 1972*a*). The present writer considers that germination is complete when all of the seed's available food reserves have been consumed and the seedling is capable of independent existence. It is the number of seeds which complete germination successfully that is of most importance to both the plant scientist and those concerned in seed technology (Chapter 9).

The germination test

The determination of the germination characteristics of a batch of seeds requires the performance of germination tests in which samples of the seeds are allowed to germinate in a petri dish or other appropriate container under controlled conditions. The germination test is so basic to an understanding of seed germination and dormancy that it requires a detailed consideration at this stage. For the main crop plants, single sets of precise conditions for the germination test for each plant have been determined and specified for use in standard tests which determine both seed viability and vigour (see Chapter 8 and International Rules for Seed Testing, 1985). For wild species and many of the less important cultivated species, the germination characteristics are either unknown or imprecisely known. In these cases, germination tests should be carried out at a number of temperatures, and the level of other parameters such as illumination, seed hydration, salinity, etc., need to be modulated if necessary.

For practical purposes, a sample of 100 seeds is most suitable for a germination test, because the number of seeds germinating can be conveniently expressed as a percentage and for each whole number between 0 and 100, the 95% confidence limits have been calculated (see Appendix) by the use of the chi-squared test and a two-by-two contingency table as described by Roberts (1963). Thus the statistical significance of the difference between two different seed samples or between similar samples from the same seed population tested under different conditions can be determined at any stage of a germination test. To reduce the 95% confidence limits given in the Appendix by half, the number of seeds required for a germination test would have to be increased to 1000. Tests at this level may be necessary for statutory but not normally for scientific purposes. In our laboratory, most of the germination tests are carried out in 90-mm-diameter plastic disposable petri dishes, initially supplied in sterile packs but washed and re-used innumerable times. For routine germination tests aseptic precautions are normally neither desirable nor necessary, as studies relate to the unsterile environment. Normal standards of cleanliness suffice. Details of the test procedures are given in Chapter 10. Work on seed biochemistry does require rigorous seed sterilization and aseptic procedures. Surface sterilization does not always eliminate microbial contamination, as micro-organisms may be incorporated deeply within the tissues. For this reason plant scientists have sometimes inadvertently studied microbial metabolism when they thought they were studying seed metabolism. In a germination test, records are normally made daily (at exactly 24 h intervals) and the results are plotted daily to give a cumulative curve (see for example Figure 1.1).

Figure 1.1 demonstrates a number of features which may be deduced from a germination test. In research publications it is usual to publish data in the form shown in Figure 1.1, although for certain purposes single parameters may be used. There is no way of calculating a single magic number which may be used to give a full description of a germination test, although the application of curve-fitting procedures can be very helpful. Germination is, for each seed, an all-or-none event, and the record of the day on which radicle emergence occurred gives the time that each seed required for this part of the germination process. Some very elementary

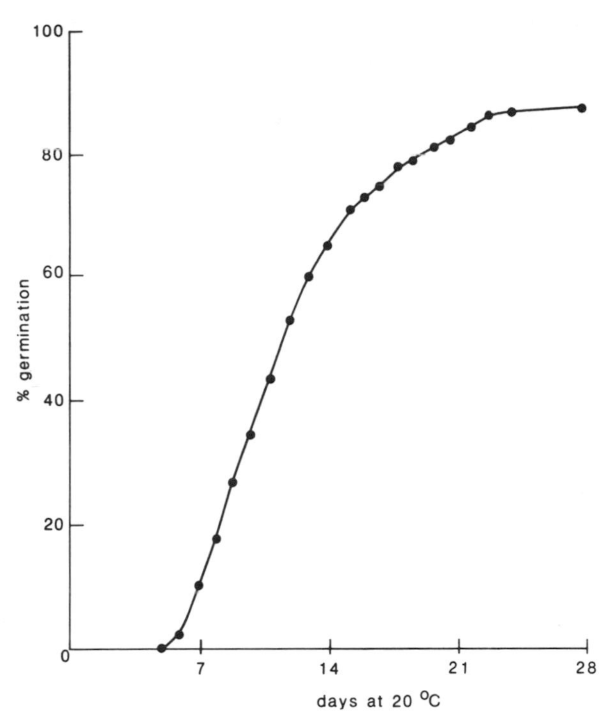

Figure 1.1 The germination of a sample of 100 intact dormant hazel seeds in 10^{-4} M GA_3 at 20 °C. After Bradbeer (1968).

deductions which may be made from Figure 1.1 are, for example, that it shows the variability in a seed sample: the two most rapidly germinating seeds achieved radicle emergence on the 6th day of the test, while the 88th and last seed to germinate during the time of the experiment germinated on the 24th day. If a seed is capable of germination it is considered to be viable. A measure of the viability of a seed sample is normally expressed as a percentage of the seeds germinating during the course of a viability test, which for the sample in Figure 1.1 is 88%. For successful establishment of crops, it is advantageous for germination to be rapid and reasonably closely synchronized. Rapidity of germination is a measure of seed vigour, which is considered, together with viability, in Chapter 8.

Dormancy

For germination to occur, seeds require moisture, a suitable temperature, and in most cases an aerobic atmosphere. If one or more of these requirements is not met, germination will fail to occur, and in this condition the seeds may be regarded as being in a state of imposed dormancy. In this book, imposed dormancy will be designated as such, and the term 'dormancy' will be used in the sense used by Wareing (1965), 'for instances where the seed of a given species fails to germinate under conditions of moisture, temperature and oxygen supply which are normally favourable for the later stages of germination and growth of that species'. Harper (1957 and 1977) said that 'some seeds are born dormant, some acquire dormancy and some have dormancy thrust upon them', and called these three categories 'innate', 'induced' and 'enforced' dormancy. Harper's 'enforced dormancy' is my imposed dormancy, but he splits dormancy as used by Wareing and myself into two categories. Innate dormancy occurs when seeds are in a dormant state on release from the parent plant, whereas induced dormancy is used to describe the situation in which dormancy develops in response to some experience after release from the parent plant. Although these two concepts are useful to the ecologist and population biologist, they can be an unnecessary complication to the plant physiologist, who often finds that a seed may possess more than one dormancy mechanism (Chapter 5) and have several ways in which dormancy can be broken (Chapter 6).

There are the terms 'relative dormancy' and 'conditional dormancy', which are synonymous, and which require to be defined at this stage. In conditional dormancy, a seed is able to germinate under a restricted range of conditions; for example, barley seeds when freshly harvested have been shown to germinate at 10 °C but not at 15 °C, but after dry storage for some time there is a widening of conditions under which germination will occur. The assessment of situations of a widening (and presumably narrowing) of the range of conditions in which germination will occur is another aspect of dormancy which has been studied particularly by Vegis (1964).

There are a number of terms in the literature of dormancy whose use seems to be inappropriate because of the misleading images that these terms produce. For example, the period during which a seed remains in the dormant state is often described as 'rest', although it is known that dormant seeds may undergo substantial or even intense metabolic activity during the dormancy-breaking processes. These dormancy-breaking processes are often termed 'after-ripening', although ripening in its primary sense implies a process in which a fruit or seed becomes ready to be reaped, gathered, or fully developed so as to reach a state suitable for consumption. I shall avoid the use of such potentially misleading terms unless the proper quotation of a literature source demands it.

CHAPTER TWO
SEED FORMATION

This chapter considers the development of seeds on the parent plant. Basic descriptions of this process can be found in elementary botany texts, and greater detail may be seen for example in Foster and Gifford (1974), in monographs such as Corner (1976), or in original papers. The main developmental patterns found in gymnosperms and angiosperms will be considered, although the many developmental peculiarities, such as those found in the Gnetales for example, will receive little mention.

Morphological aspects of seed formation

Gymnosperms

In gymnosperms, the seed originates from an ovule borne on a sporophyll. Initially the main part of the ovule is a parenchymatous tissue termed the nucellus, which is enclosed by a single integument. Integument and nucellus are fused at the base (this region is termed the chalaza), but they are separate at the top, with a small central opening, the micropyle (Figure 2.1*a*). Within the nucellus, a single megasporocyte undergoes meiosis to yield a row of four haploid megaspores. The outer cell resulting from the first of the meiotic divisions does not always undergo the second division, but in any case the innermost megaspore gives rise to the female gametophyte while the other three (or two) cells eventually degenerate (Figure 2.1*b*). The nucleus of the innermost megaspore enters a series of divisions without cell wall formation to give an enlarged coenocytic female gametophyte which in most cases contains between one thousand and several thousand nuclei. Cell wall formation occurs, commencing peripherally and proceeding centripetally, until the whole gametophyte is comprised of cells which accumulate food reserves. Cells at the micropylar end of the gametophyte produce archegonia of which the lower cell gives rise to the egg cell (Figure 2.1*c*).

Gymnosperms are usually wind-pollinated, producing large numbers of microspores (pollen) which arise from the meiotic division of microsporocytes. Subsequently, various nuclear and cell divisions occur in the

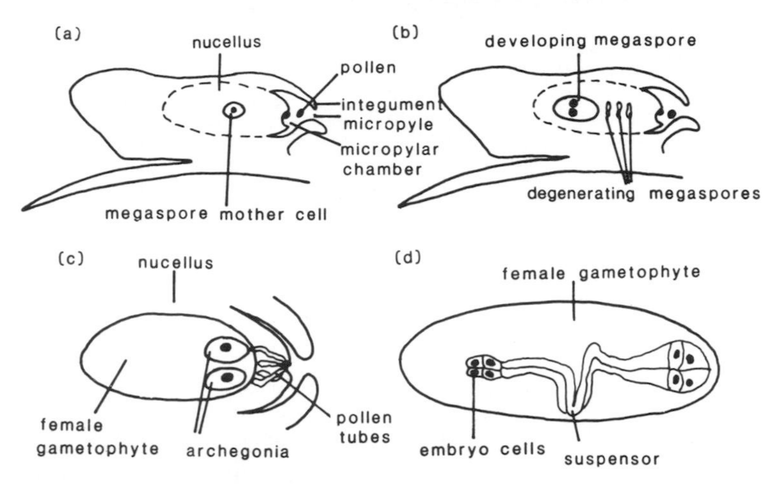

Figure 2.1 Diagrammatic representation of ovule formation and seed development in *Pinus* shown in median longitudinal section. *a*, megaspore mother cell with pollen in micropylar chamber; *b*, megaspore formation; *c*, the female gametophyte at the stage of fertilization; *d*, proembryo with suspensor and embryo cells. After Weier *et al.* (1970) and Foster and Gifford (1974).

microspore, with the eventual formation of two male gametes, one of which will fuse with the nucleus of the egg cell. Pollination involves the transfer of pollen to the micropylar end of the ovule. Most species have a long delay between pollination and fertilization, for example a year in conifers and several months in cycads and in *Ginkgo biloba*. Pollen-grain germination involves the extrusion of a tube which in cycads and in *Ginkgo* grows into the nucellus where it seems to have a haustorial function. In these plants, fertilization involves the liberation of two flagellated sperm from the pollen tube into a cavity directly above the female gametophyte. After the entry of the sperm into an archegonium, the nucleus of the sperm fuses with the nucleus of the egg cell. In conifers, the pollen tube grows through the nucellar tissue, so that the non-flagellated male gametes are carried to the archegonium where the larger male gamete fuses with the egg cell nucleus.

In almost all gymnosperms, embryo development commences by means of nuclear divisions which are at first unaccompanied by cell wall formation, a phenomenon which is very variable in extent from species to species (for example, in *Pinus*, walls begin to appear at the four-nucleus stage, but in *Dioon eduli* not until about 1000 free nuclei are present).

After cell walls are laid down, cell division continues, and the embryo becomes differentiated into a suspensor, radicle, hypocotyl, shoot apex and two or more cotyledons. The suspensor may be relatively massive as in the cycads, rather inconspicuous as in *Ginkgo*, or lengthy as in the conifers (Figure 2.1*d*). A single gametophyte may contain several embryos as a result of the fertilization of more than one archegonium and, in addition in conifers, as a result of the cleavage of a developing embryo to give four embryos. Normally all but one of the embryos are lost during the development of the seed. As shown in Figure 3.1*a*, the gymnosperm seed has a hard coat which arises from the stony layer of the integument. The food reserves are in the embryo, the female gametophyte and in the small amount of perisperm which represents the remains of the nucellus.

Angiosperms

In angiosperms, the seed originates from an ovule which occurs within an ovary. As in the gymnosperms, the ovule consists of a parenchymatous nucellus which is completely enclosed, except for the terminal micropyle, by one or two integuments (Figure 2.2). A cell within the nucellus but near

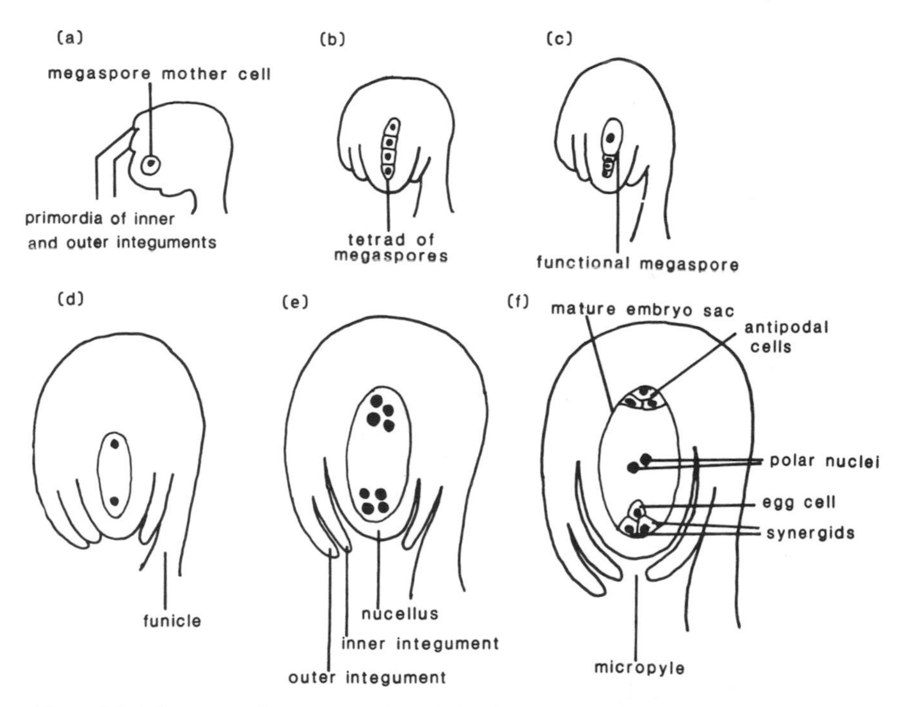

Figure 2.2 Diagrammatic representation of the development of an angiosperm ovule as shown in median longitudinal section. *a*, formation of megaspore mother cell; *b*, *c*, megaspore formation; *d*, embryo sac with two nuclei; *e*, embryo sac with 8 nuclei; *f*, mature embryo sac. After Foster and Gifford (1974).

its apex becomes a megasporocyte, which undergoes meiosis to give four haploid megaspores (Figure 2.2*a*, *b*). Some angiosperms have several megasporocytes giving rise to several tetrads of megaspores which may yield several female gametophytes, although eventually the mature seed contains only one embryo. There are two main ways in which the megasporocyte develops within the nucellus. In the tenuinucellate ovule, the megasporocyte is derived directly from a hypodermal cell in the apical region of a small nucellus. In the crassinucellate ovule, a hypodermal cell divides periclinally to give an inner cell which becomes the megasporocyte and an outer cell which undergoes further periclinal division(s) to yield up to several layers of cells. In the latter case, the megasporocyte would tend to be deeply embedded in a fairly massive nucellus.

Foster and Gifford (1974) list 11 different types of development of the embryo sac (female gametophyte). In the commonest type, described in elementary texts, the nucleus of the megaspore furthest from the micropyle undergoes three mitotic divisions to yield an embryo sac with eight haploid nuclei, three at each end and two located centrally (Figure 2.2). The nuclei at the micropylar end acquire cell walls to form the egg apparatus, comprising one egg cell and two synergid cells. The three nuclei at the end of the embryo sac furthest from the micropyle form the antipodal cells. The two centrally-located nuclei are called the polar nuclei (Figure 2.2*f*), which in one developmental type fuse to become a diploid nucleus.

Pollen grains of angiosperms are produced by microsporogenesis and have a basically simpler structure than those of gymnosperms in that when the pollen is released each grain consists of either two or three cells. The two cells are the larger vegetative cell and the smaller generative cell. The generative cell divides, either before or after pollen release, or during pollen tube extrusion, to give two male gametes. There are a range of means by which the pollen of angiosperms is transferred to the receptive stigma. The pollen tube grows down the style and enters successively the ovule and the embryo sac, where it releases the male gametes.

One of the male gametes fuses with the egg, to form a diploid zygote which gives rise to the embryo. The other male gamete fuses with one or more of the polar nuclei, the ploidy of the product of this fusion being dependent on the number of haploid genomes contributed by the polar nuclei. Most commonly the polar nuclei provide two haploid genomes, so that the fusion product is $3n$; $2n$ is also found, $5n$ is a common number and higher levels of ploidy up to $15n$ are also known.

The nucleus resulting from the fusion of male gamete and polar nuclei (nucleus) is responsible for the formation of the endosperm, which is an important feature of all angiosperm seeds other than those cases (e.g.

orchids) where the endosperm either fails to develop or degenerates at an early stage. In exalbuminous seeds, such as those of the Leguminosae, the endosperm is consumed during development, with the seed reserves being laid down in the embryo, mainly in the comparatively large cotyledons. Many species which have the bulk of their reserves in the cotyledons do possess an endosperm, frequently as one or two layers of cells surrounding the embryo, as for example in hazel and soybean. In albuminous seeds substantial amounts of reserve materials are present in the endosperm at the time of germination and the embryo may contain only relatively small amounts of reserve materials. One type of endosperm development involves free nuclear division without cell wall formation so that a multinucleate protoplasm is produced, although in some of these plants there is a subsequent formation of cell walls. In *Cocos nucifera* the liquid endosperm (coconut milk) contains free nuclei, organelles and a large number of metabolites. Coconut milk is one of the few naturally occurring liquid constituents of plants, hence its importance as a constituent of plant-cell culture media. Culture of animal cells has been aided by the occurrence of a number of potential liquid culture media in animals. The other main type of endosperm development involves the formation of cell walls immediately after each nuclear division in the endosperm.

In angiosperm embryo development, the nuclear divisions in the fusion product are in most cases followed immediately by cell wall formation, the result of these divisions being most commonly a filament of suspensor cells attached to a globular embryo as shown in Figure 2.3. Further development of the embryo results in the recognition of radicle, shoot apex and cotyledons. The typical dicotyledonous embryo develops two laterally located cotyledons which enclose between their bases the relatively small apex of the terminal shoot. Monocotyledonous embryos show a range of patterns of cotyledonary development from an apparently lateral origin of the single cotyledon to an apparently terminal origin, the origin of the shoot apex ranging conversely from terminal to lateral.

Physiological and biochemical aspects of seed formation

Although this is a subject area which has been rather neglected (Dure, 1975), during the last 10 years there has been a growing realization of its potential with respect to developmental biology and plant molecular biology, and of course for a far longer time, plant breeders and agronomists have been concerned with improvements in seed formation. There have been remarkable successes, such as the increase in average wheat yields in the United Kingdom from about 2.8 $t\,ha^{-1}$ in the early 1950s to an estimated

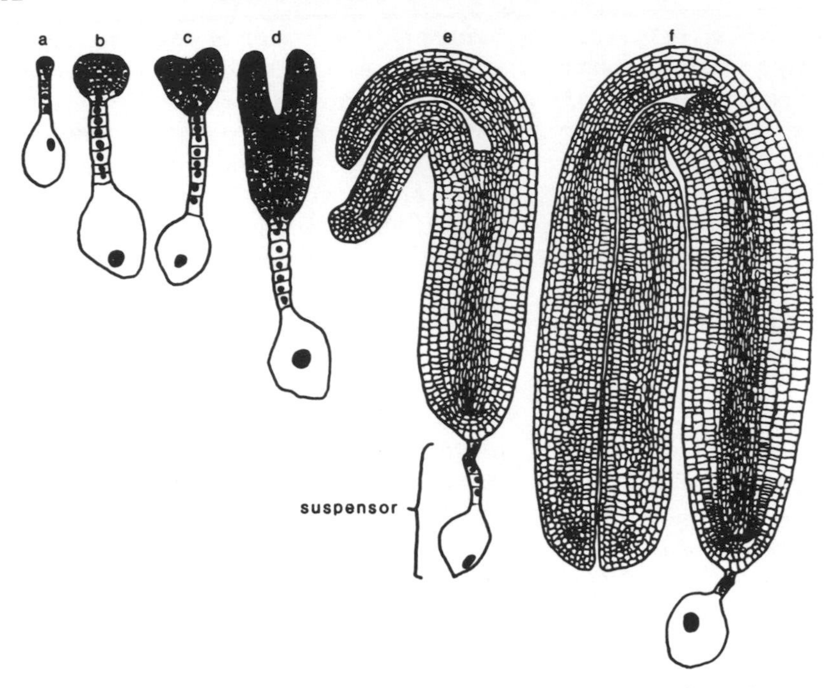

Figure 2.3 Stages of the embryogenesis of *Capsella bursa-pastoris* as seen in longitudinal section. The lower end of the embryo is directed towards the micropyle. *a*, early globular; *b*, late globular; *c*, early heart-shaped; *d*, torpedo-shaped; *e*, walking stick-shaped; *f*, fully developed. After Raghavan (1976).

record of 6.16 t ha^{-1} in 1982, an improvement which has been accompanied by an increase in the area under wheat, so that there has been an increase in United Kingdom wheat production of between 4- and 5-fold. Nor is there evidence that such increases in productivity are about to level off. On past performance, plant breeders should have the ability to stay ahead in breeding for disease resistance and, by applying the results of basic research, breeders and agronomists should continue to boost agricultural productivity.

Measurement of the course of seed development should commence from the time of fertilization, which ought to be related to some easily recognized stage of flower development. The most commonly used starting point is anthesis, which is the stage at which the pollen sacs rupture to release the pollen, a stage readily recognized in wind-pollinated plants such as grasses and sedges by the extension of the filaments of the stamens to raise the anthers above the other floral parts. Alternatively, manual pollination might be used as a starting point, or another development stage such as

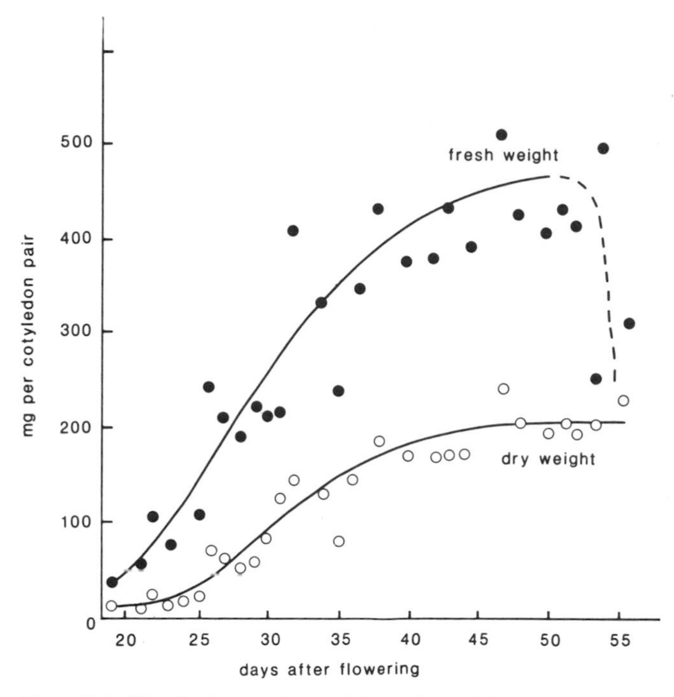

Figure 2.4 The fresh and dry weights of cotyledons of *Phaseolus vulgaris* during the development of the seeds on the parent plant. After Öpik (1968).

flower opening (in *Phaseolus vulgaris* self-fertilization occurs some 36–48 h before flower opening) or petal fall (as in *Sinapis alba*) may be used. In such work, individual flowers or simultaneously timed groups of flowers are labelled with the date on which the initial stage of the developmental process commenced.

The development of seeds and fruits or of their individual parts may be followed by measurements of physical parameters such as linear dimensions, volume, fresh or dry weights or by chemical constituents such as total N content. Figure 2.4 shows the typical sigmoidal growth curve for the fresh and dry weights of developing cotyledons of *Phaseolus vulgaris* followed by the drying out of the seed within the fruit during the later stages of ripening. Light and electron microscopy permit the measurement of many other parameters of seed and fruit formation. Perhaps the greatest difficulty in the study of plant development is that critical developmental processes appear to occur in single cells or very small groups of cells within embryos and organs such as apices of developing shoots and roots, which may contain perhaps 10^5 or more cells which may be at many different stages of many different lines of development. It is extremely difficult to recognize the cells in which the course of development of a cell line is determined; even if they could be recognized, it is most unlikely that such cells could be studied in the living state, and at present it is not possible to subject them to precise biochemical investigations.

In contrast, developing seeds do contain relatively large organs comprised of essentially homogeneous populations of cells, such as developing cotyledons or endosperm, which are very suitable for investigation. As most herbaceous plants complete their seed formation within eight weeks at the outside, and as many annuals require between 10 and 20 days for the process, it is clear that the developing seed is the site of intense synthetic activity. Knowledge of many areas of plant biochemistry has been dependent on work with immature seeds; for example, Leloir and colleagues discovered starch synthetase in immature seeds of *Phaseolus vulgaris*, knowledge of the biosynthesis of gibberellins in higher plants is based largely on investigations on immature seeds, and the first endogenous cytokinin (zeatin) was found in immature seeds of *Zea mays*.

Since we have been concerned with the study of *Phaseolus vulgaris* seedling growth, a consideration of seed formation in this species is appropriate. Investigations in the laboratories of H. Öpik (Swansea), T.C. Hall (Madison) and L.G. Briarty (Nottingham) are of particular interest, although there are many other important contributions. Comparisons between these laboratories are difficult, as each used a different cultivar, different growth conditions, and a different basis for determining stages of seed development, viz. age (Öpik), seed length (Hall) and seed fresh weight (Briarty). Strictly speaking, age is the correct basis, provided that controlled environment chambers are available (Öpik did not have these facilities) and a rigorous sampling procedure is used. The use of seed size can compensate for variation in environmental conditions and the variation in the rate of growth of individual seeds.

By light and transmission electron microscopy, Öpik (1968) demonstrated that the most noteworthy features of cotyledon development were the formation of starch grains, rough endoplasmic reticulum (RER) and protein bodies. RER function is considered to be the synthesis of polypeptides. These are secreted either from the cell, as in the case of hydrolytic enzymes secreted by the aleurone cells of germinating barley grains, or into another compartment of the cell, which for the bean cotyledon involved the secretion of newly synthesized storage proteins into the vacuoles which become protein bodies.

Briarty (1980) obtained quantitative data from electron micrographs of developing bean cotyledons by means of stereological analysis. This type of analysis may be carried out on random electron micrographs, consecutive sections not being necessary unless it is required to build up a three-dimensional model of a cell or tissue. Table 2.1 shows that, when measurements were commenced at the 20 mg seed weight, the cotyledons were at a very early stage of development, the total volume of their cells amounting to only 0.5 mm^3. However, by the 120 mg stage the total cell volume was 90 mm^3, of which 17% was plastids (amyloplasts) containing starch and 8% was RER. Starch continued to accumulate during cotyledon growth. In these cotyledons, cell division was essentially complete at the 40 mg stage, and the RER had reached its maximal development by the 120 mg stage, a 2450-fold increase of RER between the 20 mg and 120 mg stages. The storage protein synthesized by the RER was only beginning to accumulate in later stages of development studied by Briarty.

Sun *et al.* (1978) followed the accumulation of the main storage protein, which is a single globulin (globulin 1), soluble in saline but insoluble in water, and which contains three polypeptides with molecular weights of 53 000, 47 000 and 43 000. This is the main constituent of the protein bodies. Sensitive immunoelectrophoretic techniques failed to detect these polypeptides in any bean tissues other than the cotyledons, and even in the cotyledons they were not detected until 10 days after flowering. Thus the expression of the genetic information for these polypeptides is repressed throughout the development of the plant, except in the developing cotyledons, where its expression is intense. In addition to the involvement of transcription and translation in the synthesis of these polypeptides, these workers found that the translation products underwent processing by glycosylation. The understanding of such phenomena and of the way in which the genome is repressed and expressed is clearly important for the understanding of plant development. *Phaseolus vulgaris* has a seed with a relatively simple major seed storage protein—a greater number of seed storage proteins are found in most other plants.

Experimental approaches to embryogenesis

The main experimental contribution to embryogenesis has been the culture of embryos, beginning early in the present century with those from fully developed seeds, and subsequently involving the culture of younger embryos (Raghavan, 1976). Under appropriate cultural conditions, embryos of many species have been grown up to the stage at which they became capable of autotrophic growth. The culture of isolated embryos also provides a way of investigating embryo dormancy and is discussed later (pp. 48–52).

A fundamental advance in plant biology was made in F.C. Steward's

Table 2.1 Some measurements relating to the formation of starch and protein reserves in the cotyledons of developing *Phaseolus vulgaris* seeds. Measurements relating to seed fresh weight are calculated from the data of Briarty (1980) with cv. Seafarer. The globulin-1 values relating to seed fresh weight are those of Sun *et al.* (1978) for cv. Tendergreen. The 120 mg Seafarer sample is presumed to be at approximately the same developmental stage as the 10 mm long Tendergreen sample.

Developmental stage expressed as:		Amount expressed per pair of cotyledons			
Seed fresh weight (mg)	Seed length (mm)	Volume of cotyledon cells (including cell walls) (mm^3)	Volume of plastids plus starch grains (mm^3)	Volume of RER (mm^3)	Globulin (mg)
20		0.5	0.03	0.003	0.02
40		2.3	0.22	0.05	0.03
60		12.2	1.15	0.55	0.03
80		22.4	2.55	1.43	0.11
120		90.4	15.4	7.35	0.29
	10				0.2*
	12				0.7*
	14				3.2*
	16				3.8*
	18				17.2*

* Globulin-1 only.

laboratory, where a culture of free cells obtained from an explant of the secondary phloem of carrot gave rise to multicellular aggregates which went through embryo-like stages in culture before development into a mature carrot plant (Steward, 1963). This showed that a single vegetative non-meristematic plant cell had retained its totipotency, or in other words the cell possessed all of the necessary genetic information for the growth of a normal plant. Further work established that free cells of carrot were able to form large numbers of embryos which faithfully reproduced the different stages of embryogeny in culture as if they were exact replicas of zygotic embryos. Embryos arising in a vegetative manner from cells which are not themselves the immediate product of sexual union are described as adventive embryos or embryoids. Similar results have been obtained with many plants, although not all that have been tested have given positive results so far. The production of large numbers of embryoids in this way is of value in the multiplication of plants, and artificial seeds in the form of encapsulated embryoids have been produced. Embryoid production is also of importance to plant breeders in obtaining haploid plants via pollen or anther cultures and in obtaining plants from somatic zygotes resulting from the fusion of two protoplasts.

CHAPTER THREE

SEED STRUCTURE AND COMPOSITION

In a consideration of seed dormancy and germination we are concerned with the form in which a seed is released from the parent plant, whether as a naked seed or as a fruit, part of a fruit or complete inflorescence. The development of the various seed parts, namely testa, embryo, endosperm and perisperm, has been considered in the previous chapter. At the time of dehiscence, the cells of the testa, and where relevant those layers of the pericarp immediately adjoining the testa, will have normally become thickened, lignified and suberized prior to their drying and dying so as to enclose the seed in a tough protective cover. Not all seeds develop tough covering layers, however. In some seeds the testa consists of mucilage cells whose contents anchor the seed to the soil and retain the water which is necessary for germination. Other surface features, such as winged protuberances, hairs, spikes and prickles, may function variously in seed dispersal and in anchoring the seed to the soil. In those seeds where the inner layers of the pericarp take on the protective function, the testa tends to remain thin and undeveloped.

The micropyle should be detectable in all seeds, although in many seeds it is either tightly closed or sealed by a plug, the strophiole. Another feature of the seed surface is the hilum, which is the scar which remains from the point of attachment of the funicle (the stalk by which the seed was attached to the ovary wall). In seeds which developed from orthotropous ovules, the micropyle and hilum are situated at opposite ends of the seed. However, most seeds develop from a range of ovule types other than orthotropous, so that micropyle and hilum are frequently closer to each other or even adjacent.

Some seed types

The structures of a range of seeds are shown in Figures 3.1 and 3.2, and some of their dimensions are given in Table 3.1. The examples cover some of the main types of seed studied by seed physiologists.

Dehiscent seeds

These are seeds for which the pericarp of the fruit splits to release the seed. Although gymnosperms do not have a fruit, they do release naked seeds, and therefore will be treated here by a consideration of the seed of Scots pine (*Pinus sylvestris*, Figure 3.1*a*) in which the embryo comprises a whorl of 6 to 8 cotyledons which surround the shoot apex, a short hypocotyl, and a radicle. The embryo is embedded in the tissue of the female gametophyte, which in turn is enclosed by a seed coat consisting of a hard outer coat and a thin and papery inner coat, both being derived from the integument. This seed, like that of most Coniferales, is winged. Although most gymnosperms have several cotyledons, some, such as yew (*Taxus baccata*), possess only two.

The dwarf bean (*Phaseolus vulgaris*, Figure 3.1*b*) has many cultivars and has been widely used in research. The embryo consists of two large

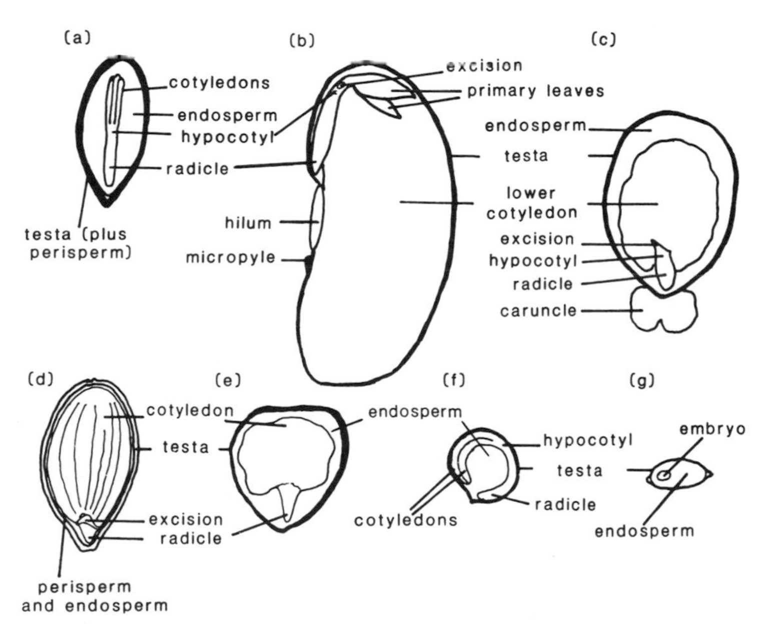

Figure 3.1 The structure of some dehiscent seeds as shown by vertically cut or dissected half seeds. Where an upper cotyledon has been excised the site of excision has been marked. *a*, *Pinus sylvestris* (Scots pine) × 4; *b*, *Phaseolus vulgaris* cv. Canadian Wonder (dwarf bean) × 2; *c*, *Ricinus communis* (castor bean) × 2; *d*, *Cucurbita pepo* cv. Long Green Bush (vegetable marrow) × 2; *e*, *Corchorus capsularis* cv. D-154 (jute) × 4; *f*, *Amaranthus viridis* × 8; *g*, *Juncus effusus* (soft rush) × 20.

cotyledons between which is found the embryonic axis comprising radicle, hypocotyl, epicotyl, two primary leaves each of about 2 mg in weight, and a shoot apex. There is no endosperm, the testa is thin, with the micropyle adjacent to the hilum and the seed does not show a dormancy mechanism.

In castor bean (*Ricinus communis*, Figure 3.1*c*) the embryo consists of two large but very thin cotyledons, a very small epicotyl, a short hypocotyl, and a radicle. It is embedded in a massive endosperm. The micropyle and hilum are close together underneath the caruncle, which is a spongy outgrowth of the testa. Castor bean has been widely used in research, chiefly with respect to the study of the mobilization of the endosperm reserves. It has the disadvantage of containing at least three highly toxic substances: a protein (ricin), an alkaloid (ricinine), and an allergen.

Vegetable marrow (*Cucurbita pepo*, Figure 3.1*d*) is of interest because the extra embryonic reserves comprise a perisperm (formed from the nucellus) of 3 or 4 layers of cells and an endosperm of a single layer of cells. The embryo consists of two thick cotyledons, a very small epicotyl, a very short hypocotyl, and a radicle. The seed is relatively flattened, and micropyle and hilum are adjacent to each other.

The two species of jute on which we have worked, *Corchorus capsularis* (included in Figure 3.1*e* and in Table 3.1) and *C. olitorius*, possess small seeds with two relatively broad cotyledons, and an embryonic axis consisting of radicle, hypocotyl and shoot apex. There is a substantial endosperm. *Amaranthus viridis* (Figure 3.1*f*), grown as a vegetable in West Africa, has a small discoid seed with a relatively substantial endosperm and an embryo curved around the periphery of the seed. The embryo consists of radicle, hypocotyl, small apical bud and two narrow cotyledons.

The one monocotyledonous seed included in Figure 3.1 is the very small seed of *Juncus effusus* (Figure 3.1*g*) which contains an endosperm in which is embedded a small rectangularly ovoid embryo which has one cotyledon and a radicle.

Indehiscent seeds

These are seeds which are not released from the pericarp of the fruit; the whole fruit, or a portion of the fruit together with the seed is released.

The Kent cobnut (*Corylus avellana*, Figure 3.2*a*) is an example of a nut in which a hard pericarp encloses the embryo. The embryo consists of two massive cotyledons which enclose at the tip a small embryonic axis consisting of epicotyl, hypocotyl, radicle and one or more leaf or scale leaf

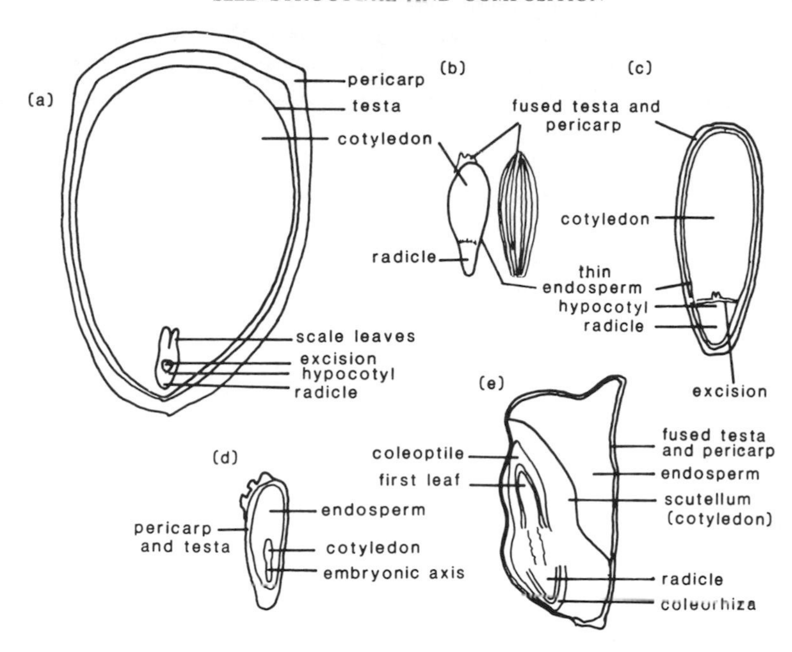

Figure 3.2 The structure of some indehiscent seeds as shown by vertically cut or dissected half fruits. Where an upper cotyledon has been excised, the site of excision has been marked. *a, Corylus avellana* (Kent cobnut) × 2; *b, Lactuca sativa* cv. Webbs Wonderful (lettuce) × 4, upper testa and pericarp dissected from seed; *c, Helianthus annuus* (sunflower) × 2; *d, Daucus carota* (carrot) × 4; *e, Zea mays* cv. Earliking (sweet corn) × 4.

initials. There is a very thin endosperm, amounting to one or two cell layers in thickness, enveloping the embryo, and a thin papery testa. Although the embryo of the newly harvested hazel nut is not dormant, dormancy develops as the embryo dries.

The seed of lettuce (*Lactuca sativa*, Figure 3.2*b*) has been used for the investigation of dormancy in those cultivars which are photoblastic above 25 °C. As in all members of the Compositae, each seed is a single-seeded fruit, the cypsela, in which the fairly thin pericarp is fused to the testa. Within the seed is a fairly thin endosperm, and an embryo which includes a radicle, hypocotyl, shoot apex and two narrow cotyledons. Sunflower (*Helianthus annuus*, Figure 3.2*c*) is another member of the Compositae, but with a much larger seed than lettuce, with broader cotyledons and a substantial endosperm.

In the Umbelliferae, the fruit is a schizocarp which consists of two dry one-seeded carpels which split along the midline on release to give single-seeded halves in which testa and pericarp are fused. Carrot (*Daucus carota*) is the example shown in Figure 3.2*d*. The embryo consists of hypocotyl, radicle, two small cotyledons and an apical bud, and the main reserves are in the endosperm.

The seed of cereals and grasses has the testa fused to the pericarp to give a fruit described as a caryopsis. The example shown in Figure 3.2*e* is corn (*Zea mays*). All members of this group contain an embryo with a single cotyledon, the scutellum, which acts as an absorbing organ during germination when the reserves in the endosperm are hydrolysed. The endosperm is normally hard and flinty, as the cells die once the reserves, mainly carbohydrate, are laid down. The endosperm is enclosed by the aleurone layer, which usually amounts to several layers of living cells with high amounts of reserve protein.

Seed size

The seeds included in Table 3.1 show a range in size of between almost 2g for *Corylus avellana* and 10 μg for *Juncus effusus*, a range of over five orders of magnitude. To my knowledge, the largest known seed is that of the

Table 3.1 Some physical measurements of the seeds described in the text and illustrated in Figures 3.1 and 3.2.

Species	Seed length in mm	Weight in mg		
		Whole seed	Embryo	Extra-embryonal reserves
Pinus sylvestris	4.0	6.8	0.3	5.0
Phaseolus vulgaris cv. Canadian Wonder	15	401	369	—
Ricinus communis	12	325	8	236
Cucurbita pepo cv. Long Green Bush	13	95	78	4
Corchorus capsularis cv. D-154	2.8	3.6	0.8[1]	1.7[1]
Amaranthus viridis	1.1	0.35	0.1	0.15
Juncus effusus	0.4	0.01	0.001[1]	0.007[1]
Corylus avellana	22	1450	1400	20[1]
Lactuca sativa cv. Webb's Wonderful	4.1	1.2	0.97	0.1
Helianthus annuus	16	93	51	4
Daucus carota	3.3	2.0	0.2[1]	1.5[1]
Zea mays cv. Earliking	8	208	58	133

[1]Calculated from volume.

double coconut (*Lodoicea sechellarum*), in which the fruit weighs about 20 kg and contains a single seed of matching size. Thus the known range of seed size is just over 9 orders of magnitude. Clearly, in the upper part of the range seeds have the capacity to contain large embryos and substantial food reserves which would enable a seedling to achieve considerable growth of both root and shoot before it became dependent on its own photosynthesis, or, in the case of parasitic angiosperms, on an appropriate host plant. Salisbury (1942) points out that in the extreme development of this strategy in the double coconut, the large size of the fruits is associated with a low rate of fruit production. The palm bears only 4–11 fruits at a time, each fruit takes 10 years to develop, and female trees do not bear fruit until they are 30 years old. The fruit had been found at sea and washed ashore long before the discovery of the location of the adult trees on two islands of the Seychelles group. The seed is apparently unable to maintain its viability during dispersal by ocean currents, but despite the low reproductive rate it flourishes in its natural habitat, where the competitive ability of its very large seedling is presumed to be advantageous.

To state the obvious with respect to seed production, more seeds will mean smaller seeds, and alternatively, in order to obtain larger seeds a smaller number of seeds will be produced. To maintain its abundance in the environment each individual of a species must give rise, on average, to just one daughter (presumably two for female plants of dioecious species with a 1:1 male:female ratio) which itself achieves the reproductive stage. Salisbury's monograph, *The Reproductive Capacity of Plants* (1942), deals with these matters at length. He concluded that

> for most species at least the seed output is considerably in excess of that requisite for mere replacement of losses by death and sufficiently so to bear no obvious relation to normal seedling mortality. The size of the seed output, or more precisely the reproductive capacity, is regarded as a positive asset in the competitive equipment of the species which tends to ensure occupancy of the available ecological niches and so to increase the species' frequency and distribution.

If the successful establishment of a seedling is dependent on a seed reaching a specialized niche then the production of large numbers of small seeds should increase both the range of dispersal and the chances of the successful occupation of an appropriate niche. Salisbury points out that minute seeds (about 5 μg per seed) are especially characteristic of certain groups such as those whose growth is dependent on an association with a mycorrhizal fungus (Orchidaceae, Ericaceae and Pyrolaceae), on a parasitic association with another higher plant (Orobanchaceae), or in wholly saprophytic species which show little or no photosynthesis (e.g. some members of the Orchidaceae). Successful establishment of seedlings from these plants is dependent on the germination of the seed in immediate proximity to the appropriate fungus or higher-plant root, as the seed contains little reserve material. There are also meagre amounts of reserve materials in the minute seeds of some autotrophic plants whose characteristic is to germinate in open but moist habitats (e.g. *Drosera* spp. and *Juncus effusus*.

Seed reserve substances

Microscopic examination of a mature seed shows that all of the cells of the perisperm, endosperm and embryo have the appearance of being full of the reserves upon which the germination of the seedling is dependent up to the point of its becoming photosynthetically self-sufficient or its achieving access to some other nutrient supply.

The main reserves, represented by lipids and carbohydrates, provide carbon skeletons and energy (in the form of high-energy phosphate bonds and reducing power) for embryo growth and all of the processes involved therein. Although most of the seeds which are used as major human foods contain starch as their main reserve, of the seeds so far investigated, about 90% of species have lipids as their main seed reserve. The lipid reserves occur as oils (liquid at ambient temperatures) being the neutral esters of fatty acids and glycerol. Oil seeds are of considerable commercial importance for the extraction of edible oils and of oils for other purposes, the residue or cake left after extraction normally being used as an animal feedstuff. Vaughan (1970) described oil seeds of commercial importance from 48 angiosperm families as well as 7 gymnosperm seeds. The oil occurs within the cells as oil droplets of varying size. The starch-containing seeds which contribute to the human diet are mainly cereal grains and pulses, starch being their major component and lipids a minor component. The starch grains form in modified plastids, amyloplasts, the envelopes of which will frequently be found to have ruptured during the drying of the seed in ripening. Other reserve carbohydrates such as hemicellulose have been found in some seeds.

Storage proteins represent an important reserve substance in all seeds, and they provide a major component of the dietary proteins for most humans and many domesticated animals. In seeds the hydrolysis of reserve proteins during germination provides the amino acids for the synthesis of new proteins in the germinating seedling.

Seeds contain sufficient mineral elements for seedling growth. Phosphorus is stored as phytate which is inositol hexaphosphate. The synthesis of the reserve proteins occurs on an extensive RER system, some of whose ribosomes and other nucleic acids may be used directly in the course of the synthesis of the proteins required during germination. In addition, some of the nucleic acids are recycled during germination.

An interesting class of compounds often represented in seeds is that of secondary plant products containing nitrogen. A secondary plant product is a substance formed at the end of a secondary chain of biosynthesis which is not utilized by the plant for any other biosynthetic purpose. Some of these compounds are non-protein amino acids, which means amino acids other than the 20 which occur naturally in proteins. Some 300 non-protein amino acids have been characterized in plants and have normally been found to reach their highest concentrations in seeds. Other types of secondary plant products, such as amines, alkaloids and cyanogenic glycosides, also accumulate in seeds. Many such substances reach high concentrations of up to 10% or more of the dry weight. It is not possible to generalize as to whether they are metabolized during germination, for in some cases, such as that of *Medicago sativa*, canavanine in the seed is rapidly metabolized during germination and growth, whereas in other cases, such as that of *Albizzia julibrissin*, there was no change in the seed content of albizziine during germination and seedling growth. Mobilization of such molecules involves their utilization as reserves of nitrogen, of carbon skeletons and in some cases of sulphur, but may also involve their exudation from the roots, in which they may have an allelopathic role in inhibiting the growth of other plants or of micro-organisms.

Most of these secondary compounds appear to offer a chemical defence against predators and microbial pathogens. In its most effective form, a chemical defence deters all predators by smell, taste or appearance. A less effective defence will kill, injure or otherwise interfere with the development of a predator, and thus reduce the amount of predation. Chemical defence mechanisms are often breached by specifically adapted predators, so that a seed may provide the sole food source for a predator. For example, the larvae of the bruchid beetle *Caryedes brasilensis* in Costa Rica feeds exclusively on seeds of *Dioclea megacarpa*, which contain canavanine to the extent of 7–10% of the fresh weight of the seed. The beetle avoids incorporating the canavanine into its proteins, and detoxifies it prior to utilizing its nitrogen (Harborne, 1982). The natural resistance of seeds to fungal and bacterial attack is also considered to be, in part, an aspect of chemical defence. The presence of secondary compounds in seeds may make the seed toxic to man, as for example in the case of neurolathyrism which results from the consumption of an excessive amount of *Lathyrus* meal (about 50% of the diet), particularly in time of famine in the Indian subcontinent. The substance responsible is α-amino-β-oxalyl-amino-propionic acid.

More information about the chemical constituents of seeds can be found in texts on plant biochemistry (Goodwin and Mercer, 1982), seeds (Bewley and Black, 1978) and ecological biochemistry (Harborne, 1982).

CHAPTER FOUR

GERMINATION

In this chapter consideration will be given to the course of germination up to the stage of seedling establishment. For green plants, from which all the examples are taken, successful seedling establishment means that the seedling is sufficiently intact to have the expectation of reaching maturity, that it has a root system sufficiently established to provide it with appropriate anchorage, water and mineral salts, and that the development of its photosynthetic organs has resulted in the seedling having begun to make a net gain in dry weight.

The physical requirements for germination

The conditions required for the germination of dormant seeds (in true dormancy as defined on p. 6) are discussed at length in Chapter 6. Here attention will be paid only to the requirements for the germination of non-dormant seeds, i.e. those that germinate under conditions which are suitable for the growth of the mature plant. The exact range of requirements for germination is peculiar to each species and even to each seed sample.

(i) *Water*. All seeds require sufficient moisture for imbibition and germination and most will germinate well even when supplied with a clear excess of water. For some seeds, such as beetroot (*Beta vulgaris*) and spinach (*Spinacea oleracea*), excessive amounts of water reduce the permeability of the coat to oxygen and inhibit germination (Gulliver and Heydecker, 1973).

(ii) *Temperature*. Each species has a range of temperatures at which germination will occur and at which seedling establishment is possible. At unsuitably low temperature, imbibition may be possible, but either no embryo growth may follow, or low-temperature damage to embryos or seedlings may prevent the completion of germination. At marginally low temperature, photoinhibition of photosynthesis may take the form of a bleaching of the photosynthetic membranes. Over-enthusiasm of amateur gardeners in sowing or planting too early in the season can lead to disappointment from low-temperature damage, which can occur above freezing point. Similarly, unsuitably high temperatures may permit imbibition but not permit embryo growth or seedling establishment. Feierabend and co-workers discovered that, for a number of crop plants, at certain critical high temperatures seedling growth occurred, but the seedlings failed to become green. The apparent reason is that the plastid ribosomes fail to assemble at the non-permissive high temperature, and consequently no proteins are synthesized within the developing plastid (Feierabend, 1979). Chloroplasts possess DNA which codes for about 25% of their constituent polypeptides and which are synthesized on plastid ribosomes, the remainder of the chloroplast polypeptides being coded by nuclear DNA and synthesized on cytoplasmic ribosomes (Bradbeer, 1981). At the critical temperature, cytoplasmically synthesized proteins accumulate in the plastids but photosynthetic membranes and pigments are not formed. However, a lowering of the temperature for as little as one hour permits assembly of the plastid ribosomes so that normal greening can take place even on return to the non-permissive temperature. The non-permissive temperatures for the crop plants which were examined were close to 34°C.

(iii) *Oxygen.* Although imbibition will occur in the absence of oxygen, for most seeds the remaining stages of germination are inhibited under anaerobic conditions, and if anaerobiosis persists, seed death results. Oxygen is the terminal electron acceptor in respiration, and an absence or an insufficiency of the oxygen supply inhibits the respiration necessary for the germination of most seeds and also results in an accumulation of potentially toxic products of anaerobic respiration, such as acetaldehyde, ethanol and lactate. Seeds adapted to germination under water, such as rice (*Oryza sativa*), tend to be able to grow under anaerobic conditions. Except in such cases, it is difficult to recognize situations in nature where a deficiency of oxygen within the tissues of a non-dormant seed is likely to inhibit germination.

The morphology of seed germination

Phaseolus vulgaris (Figure 4.1). In the first hours of imbibition the testa becomes wrinkled, the embryo begins to swell and the testa shows sings of splitting. Between 48 and 72 h after the beginning of imbibition, radicle emerges and begins to grow downwards into the soil, and at a comparatively early stage secondary roots are formed. On the sixth day, hypocotyl

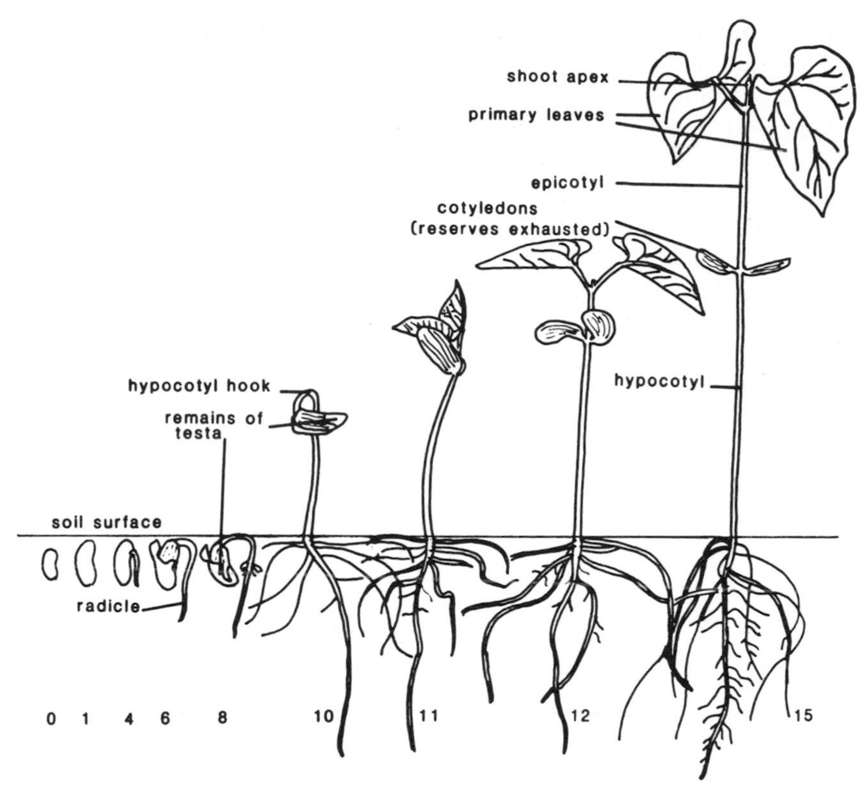

Figure 4.1 The course of germination and seedling growth of dwarf bean (*Phaseolus vulgaris* cv. Canadian Wonder) in a Conviron growth cabinet with a diurnal cycle of 12h light (63 μE.m^{-2}.s^{-1}) at 23°C and 12h darkness at 15°C. Stages shown after 0, 1, 4, 6, 8, 10, 11, 12 and 15 days. Magnification × 0.3. Note that the rate of germination and seedling development is temperature-dependent.

growth is obvious, and the hypocotyl hook forces itself to the surface of the soil on the eighth day. The hypocotyl then carries the cotyledons above the surface of the soil, it elongates rapidly and straightens out. The cotyledons open out, the remains of the testa normally fall off at this stage, and the cotyledons become green. At this stage the epicotyl becomes obvious; it is in the form of a hook and it bears the primary leaves which, up to this stage, had been protected and enclosed by the cotyledons. Under illumination the epicotyl lengthens and straightens, and the two primary leaves unfold and are carried up to lie at 90°C to the incident illumination. The primary leaves have substantial petioles. The apical bud lies between the bases of these petioles and its growth leads to the further development of the shoot. As the cotyledons are carried above the soil, the germination of bean is described as epigeal.

Zea mays (Figure 4.2) Imbibition results in the expansion of both embryo and endosperm, and the testa is penetrated by the coleorhiza (root sheath) on the fourth day. The primary root quickly emerges, to be followed by the coleoptile and the secondary roots. The coleoptile emerges from the soil and grows rapidly for a time before its growth ceases and it becomes penetrated by the first and then the second leaves of the seedling. The leaves emerge from the coleoptile with the lamina rolled up, and in response to illumination the leaves unroll in a phytochrome-mediated response. As the cotyledon remains in the soil, maize germination is described as hypogeal.

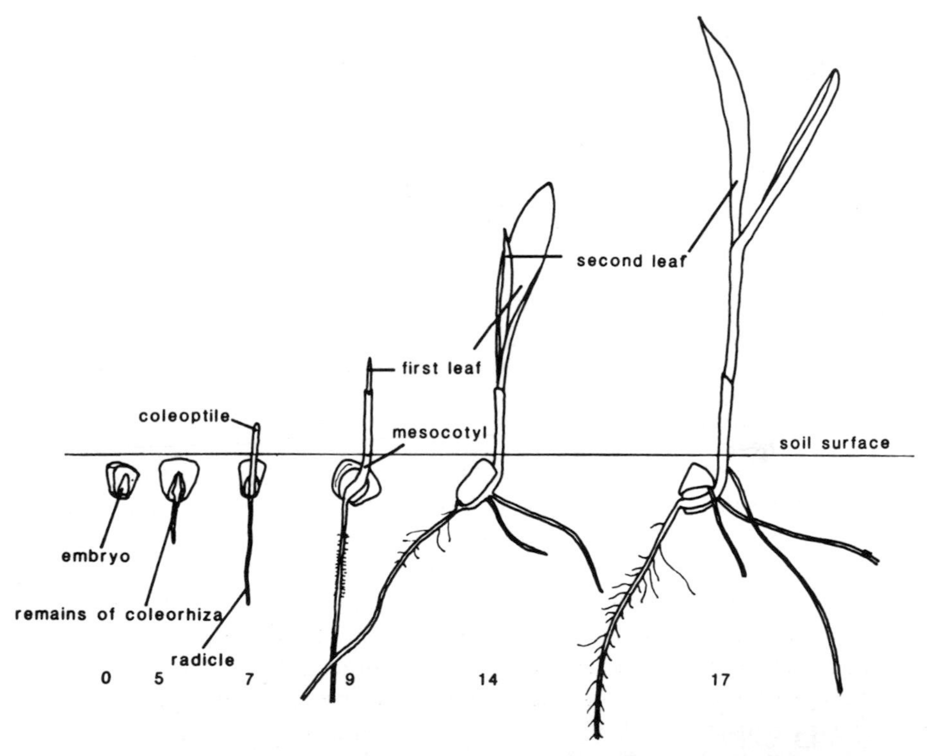

Figure 4.2 The course of germination and seedling growth of *Zea mays* cv. Earliking under the same diurnal cycle as that described under Figure 4.1. Stages shown after 0, 5, 7, 9, 14 and 17 days. Magnification × 0.5.

Comparative morphology of seedlings

Figure 4.3 shows some germinated seedlings which may be compared with the equivalent germination stages in Figures 4.1 and 4.2. The figure is largely self-explanatory, *Pinus sylvestris* showing the whorl of cotyledons typical of gymnosperms, *Corylus avellana* having large cotyledons with hypogeal germination, and the small seeds of *Amaranthus viridis* and *Corchorus capsularis* showing epigeal germination with narrow strap-shaped and relatively broad cotyledons respectively. *Allium cepa* (onion) possesses a small seed from which radicle and cotyledon emerge on germination, with only the tip of the cotyledon remaining within the endosperm of the seed, where it functions as an absorbing organ, while the remainder of the cotyledon is green and has a photosynthetic role. The seed

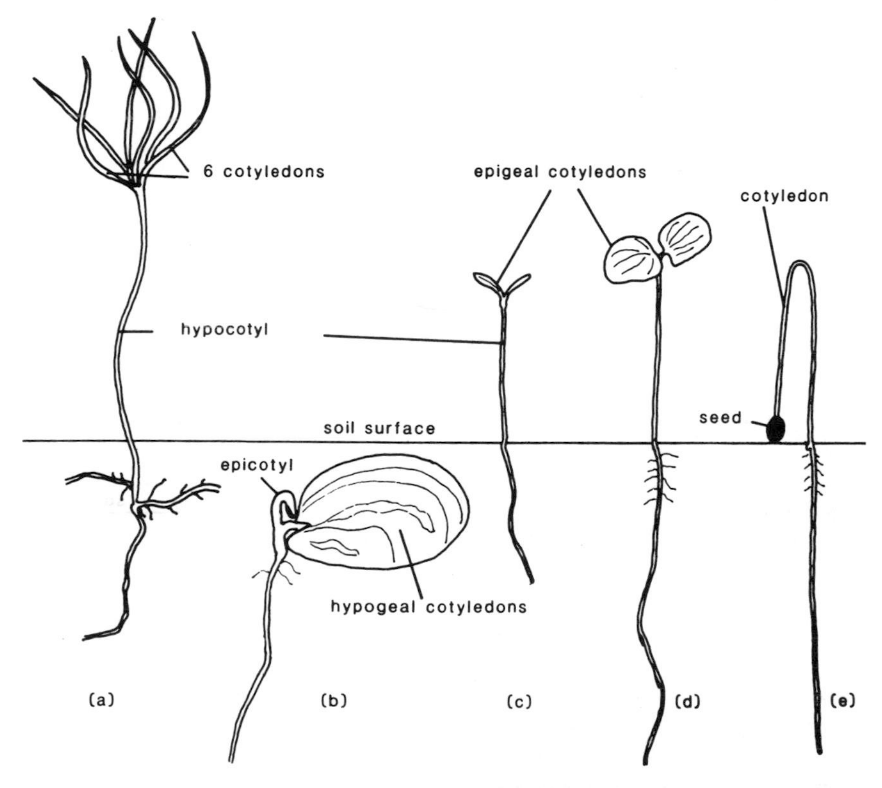

Figure 4.3 Seedlings after germination and growth in cycles of 12 h light : 12 h darkness. Actual size. *a*, *Pinus sylvestris* after 28 days at 20 °C; *b*, *Corylus avellana* (hazel) after 14 days at 20 °C; *c*, *Amaranthus viridis* and *d*, *Corchorus capsularis* (jute) after 4 days at 30 °C; and *e*, *Allium cepa* (onion) after 11 days in a 23°C :15°C cycle.

is carried above the surface of the soil, adventitious roots develop, and eventually the plumule emerges from the ensheathing cotyledon.

The physiology of germination

Imbibition

When a seed is provided with water, the water is taken up by imbibitional forces within the seed. In the air-dry seed, membranes have lost some of their integrity with the loss of water. During imbibition, membranes reassemble themselves, although some leakage of solutes occurs before membrane integrity is fully restored. Figure 4.4 shows that both water uptake and respiratory oxygen uptake of *Phaseolus vulgaris* cotyledons increase substantially during the first 12 h after the commencement of imbibition, then water uptake continues at a lower rate for up to 6 days, oxygen uptake remains steady until 24 h and then a further rise in respiration commences. The process of imbibition is reversible during the first imbibition phase (about 12 h) as shown by the ability of 6-hour imbibed seeds to be repeatedly dried and rehydrated without any loss of

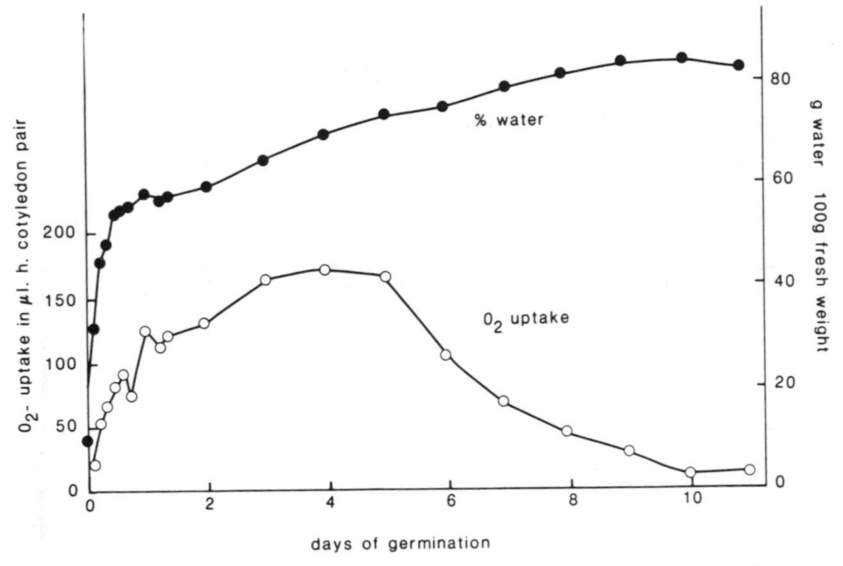

Figure 4.4 The water content and oxygen uptake of the cotyledons of *Phaseolus vulgaris* cv. Belfast New Stringless during imbibition of the seeds at 25°C in darkness. After Öpik and Simon (1963).

viability (Öpik and Simon, 1963). During this phase, the rate of oxygen uptake is proportional to the moisture content of the cotyledons. Imbibition is a physical process whose rate is not affected by temperature over the physiological range (approximately 0–40 °C). Imbibition is also able to take place when dry seeds are placed in fairly concentrated solutions of impermeable solutes whose water potentials are too low to permit embryo growth. Some recent developments in seed technology, which are described in Chapter 9, involve the manipulation of seed imbibition in pre-sowing treatments.

Growth and seedling establishment

Once growth of the root and shoot meristems has commenced, the process of germination cannot be reversed by dehydration without bringing about seedling death. Radicle and then plumule growth commence, in both cases involving cell elongation and division at an early stage and later cell enlargement, differentiation and maturation. A great deal of plant physiological research has been carried out on germinating seedlings with the objective of gaining an understanding of growth and development, of the internal mechanisms such as those involving plant growth substances which regulate these processes, and of the mechanisms by which growth and development are regulated by environmental factors. Seedlings are also useful for the study of the basic physiology and biochemistry of plants, as seedlings grown under standard conditions permit the provision of large numbers of almost identical plants for research purposes.

The measurement of growth parameters such as fresh weight, dry weight, stem height or the area of a specific leaf, will, under controlled environmental conditions, give sigmoidal growth curves similar to those shown in Figure 4.5.

Etiolation

When seedlings are grown in total darkness they develop the characteristics of etiolation, namely a tall, weak, straggling stem, very small unexpanded leaves and a yellowish-white colour. As an example the 12-day-old dark-grown dwarf bean in Figure 4.6 should be compared to the adjacent plant grown at the same temperature in 12 h days. The syndrome of etiolation is an adaptation by which the germinating seedling attempts to raise its photosynthetic organs above the soil, leaf litter and shading vegetation to

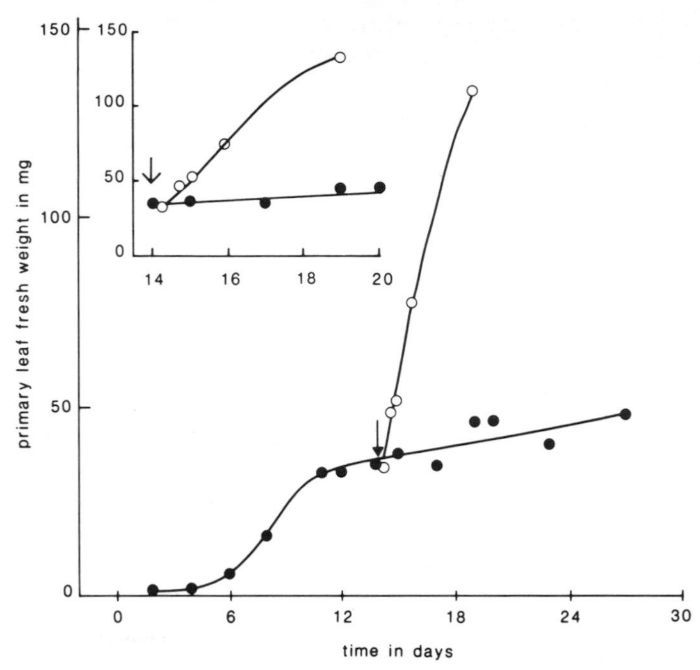

Figure 4.5 The fresh weight of the primary leaf of *Phaseolus vulgaries* cv. Alabaster during the course of germination in darkness at 23°C (●) and also after transfer to continuous fluorescent illumination of 63 μE.m^{-2}.s^{-1} after 14 days' dark growth (↓). Inset: effect of illumination. Results of J.W. Bradbeer, J. Smith and J. Rest.

reach a level of irradiation sufficient to maintain the seedling in photosynthetic autonomy. Illumination prevents the development of the various symptoms of etiolation, and in etiolated seedlings brings about de-etiolation, mainly through the photoreceptor phytochrome, a light-sensitive chromoprotein which is considered to be responsible for most of the photomorphogenetic responses of higher plants. Photomorphogenesis is the process by which the normal development of the plant occurs in response to illumination. In de-etiolation there are a number of distinct phytochrome-regulated processes; some are negative effects, such as the inhibition of shoot growth, and some are positive, such as leaf expansion and the development of the photosynthetic apparatus. Other photomorphogenetic photoreceptors involved in de-etiolation are cryptochrome (sensitive only to blue irradiation), protochlorophyllide, and the photosynthetic machinery itself (Bradbeer, 1981).

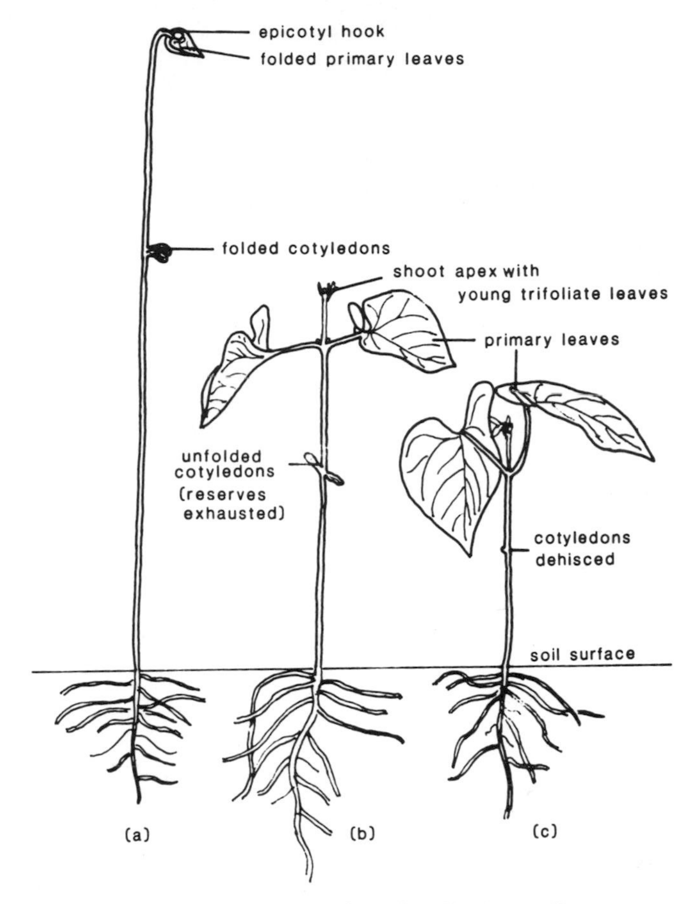

Figure 4.6 *Phaseolus vulgaris* cv. Canadian Wonder seedlings grown for 14 days at 23°C under. *a*, continuous darkness; *b*, 12 h days under fluorescent and incandescent illumination of 63 $\mu E.m^{-2}.s^{-1}$, and *c*, continuous illumination of 63 $\mu E.m^{-2}.s^{-1}$. Magnification × 0.2.

Development of the photosynthetic machinery

Seedling survival is dependent on a fairly prompt development of the photosynthetic machinery so that photosynthetic autonomy is achieved rapidly. A net CO_2 fixation may not be reached on the first day that a developing leaf is illuminated, and the growing seedling will probably take several days to achieve an overall carbon accumulation. In dicotyledonous leaves the development of the chloroplasts is usually accompanied by leaf expansion, cell division, stomatal formation, the development of a system of intercellular spaces and chloroplast division. In monocotyledonous leaves, chloroplast development is similar to that in dicotyledons, but less cell division and cell enlargement occur when the leaves unfold or unroll. For further details about chloroplast biogenesis, see Kirk and Tilney-Bassett (1978) and Bradbeer (1981).

The biochemistry of germination

Mobilization of reserves

In the embryo, reserves are stored in virtually all living cells, and it seems that they are mobilized by enzymes which are synthesized in the same cells. When Öpik (1966) studied fine-structural features of *Phaseolus vulgaris* cotyledons during germination, she found that digestion of reserves began in cells furthest away from the vascular bundles and the epidermis. From four days after imbibition these central regions were found to contain dead, empty cells, while the cells around the vascular bundle and the epidermis were still packed with reserves. Starch grains and protein bodies were digested more or less simultaneously in the same cell, and lipid reserves in the form of oil droplets disappeared. At the beginning of the digestion of the protein bodies there was considerable development of an endoplasmic reticulum bearing polysomes in the form of spirally arranged ribosomes. Digestion of the reserves was preceded by the development of mitochondria, and the later stages of digestion saw mitochondrial breakdown (Öpik, 1965). In the epidermis and subepidermal layers, chloroplasts developed in light. However, the cotyledons shrivelled and dropped some 12–14 days after imbibition.

A similar course of events occurs in the endosperm of those seeds in which the cells are living. In castor bean (*Ricinus communis*) endosperm, the main reserve material is lipid (60% of the endosperm dry weight is triglyceride), which is mainly converted into sucrose, the sucrose being absorbed by the cotyledons, after which most was translocated to the embryonic axis (Kriedemann and Beevers, 1967). The whole complex mechanism for the conversion of triglyceride to sucrose develops in the endosperm itself, and involves enzyme synthesis and the production of an array of intracellular organelles (including ribosomes, rough endoplasmic reticulum, mitochondria, plastids and glyoxysomes), much of this development occurring at the expense of the protein reserves in the endosperm of the ungerminated seed (Beevers, 1975).

In contrast, the cells of the endosperm of cereals and grasses which contain starch are dead. The starch is hydrolysed by α-amylase secreted by the aleurone cells, a layer three or four cells deep surrounding the remainder of the endosperm and containing reserves of protein and lipid (see Figure 3.2). In barley seed, in response to gibberellins secreted by the embryo, a synthesis *de novo* of α-amylase occurs in the aleurone cells, the enzyme is secreted from the cells, it hydrolyses the starch of the endosperm, and the sugars eventually produced are absorbed by the scutellum for utilization in embryo growth. A concise account of the elucidation of the main features of this process has been provided by Jones and Stoddart (1977). Some time (6 h or more) after gibberellin treatment, barley aleurone tissue releases a number of hydrolytic enzymes including α-amylase, protease, phosphatase, β-glucanase, ribonuclease, pentosanase, peroxidase, esterase and glucosidase. In a study of proteins released from aleurone layers in response to gibberellins, Jacobsen and Knox (1974) found that 10 of the proteins contained newly synthesized components and two did not, the latter apparently having been produced by the proteolysis of a precursor protein. The scutellum absorbs the products of the hydrolysis of the other endosperm materials as well as the sugars produced from starch.

Protein synthesis

Protein synthesis plays an important part in germination, in the growth of the embryonic axis and in the synthesis of the hydrolytic enzymes and the other cellular machinery used for the mobilization of the reserves. Protein synthesis can be shown to occur within a few hours of the beginning of imbibition before the recommencement of mRNA synthesis. Dry seeds contain two classes of mRNA: residual mRNA which codes for proteins synthesized during embryogenesis, and stored mRNA which was synthesized during embryogenesis but is translated during germination to yield proteins which are required for the germination process. Some understanding of protein synthesis during embryogenesis and germination has been obtained from intensive studies on cotton embryos (Dure *et al.*, 1979). As germination progresses, the synthesis of new mRNA commences and additional proteins are synthesized.

CHAPTER FIVE

THE MECHANISMS OF SEED DORMANCY

In this chapter and the next, the phenomenon of dormancy, as defined on p. 6 and excluding imposed dormancy, will be considered, first of all with respect to the mechanisms of dormancy and secondly by discussing the ways in which dormancy may be broken. Since it has been by studying the breaking of dormancy that much of our understanding of the mechanisms of dormancy has been obtained, this order of treatment is not wholly logical. Furthermore, at a time when new advances in molecular biology are announced almost daily, we have to confess that our understanding of the molecular basis of seed dormancy is so fragmentary that there is no plant species for which there exists an adequate account of its dormancy in cellular and molecular terms. The most comprehensive survey of the literature of seed dormancy may be found in the second volume of the monograph by Bewley and Black (1982), in which about 400 species are listed for which substantial research contributions on seed dormancy have been made, but the authors point out that they have omitted very many contributions and presumably many species. Table 5.1 lists the species and genera whose seed dormancy seems to have received most attention from research workers. In addition to those species in which dormancy has been studied, there are many more species which are known or reliably suspected to possess dormant seeds. Where seed germination has been studied in a

Table 5.1 The species and genera whose seed dormancy has received most attention by research workers. The list in declining order of interest, is based on the numbers of papers published in major journals

Lactuca sativa	Lettuce
Malus sylvestris	Apple
Fraxinus spp.	Ash
Xanthium pennsylvanicum	Cocklebur
Acer spp.	Maple, sycamore
Corylus avellana	Hazel
Euonymus europaea	Spindle tree
Avena fatua, A. ludoviciana	Wild oat
Rumex spp.	Dock

Table 5.2 The main identified mechanisms of seed dormancy

A. Dormancy caused by the embryo coverings (pericarp, testa, perisperm and endosperm)
   1. Restriction of gaseous exchange
   2. Restriction of water uptake
   3. Mechanical restriction of embryo growth
   4. Water-soluble inhibitors in the embryo coverings
   5. Dormancy from the failure to mobilize extra-embryonic food reserves

B. Embryo dormancy
   1. Underdeveloped and undifferentiated embryos
   2. Block to nucleic acid and protein synthesis
   3. Failure to mobilize food reserves of embryo
   4. Deficiency of plant growth substances
   5. Presence of inhibitors

range of wild species, it has been found that the majority of species show seed dormancy (see Chapter 7).

In Table 5.2 I have attempted to list the main dormancy mechanisms for which evidence has been presented. It should be noted that a number of workers have previously published classifications of the mechanisms of seed dormancy. The most noteworthy accounts are by Barton (1965*a*), Nikolaeva (1967, 1977), Villiers (1972), Harper (1977), Bewley and Black (1982) and Mayer and Poljakoff-Mayber (1982).

Table 5.2 shows that dormancy mechanisms may reside in two distinct sites, namely the embryo coverings and the embryo itself. Should a dormant seed contain an embryo which proves to be capable of germination after isolation, it may be concluded that the cause of dormancy resides in the layers which enclose the embryo. However, many dormant seeds possess more than one mechanism of dormancy. Therefore it must not be assumed that if a dormant seed contains a dormant embryo the seed coverings may not contribute to the dormancy of the seed.

Dormancy caused by embryo coverings

The coverings which are part of the seed are the endosperm, the perisperm when present, and the testa, although endosperm and perisperm do not always completely enclose the embryo. However, since many seeds are not released from the fruit in nature, it is necessary to consider the whole dispersal unit rather than the seed *sensu stricto*, and to include, where appropriate, the embryo coverings arising from the pericarp and the other persistent fruit, flower and inflorescence parts.

The restriction of gaseous exchange by the embryo coverings

There is no doubt that embryo coverings provide a barrier to gaseous exchange, although it has yet to be proved that such a limitation to gaseous exchange can cause a seed to be dormant. Of relevance are investigations on the dimorphic seeds of *Xanthium pennsylvanicum*, of which the smaller upper seed in the capsule is dormant whereas the larger lower seed is not. Circumstantial evidence convinced a number of workers that the seed coat was responsible for dormancy in this case (for references see Porter and Wareing, 1974). This evidence was that the isolated embryo of the dormant upper seed is able to germinate, that this embryo has a greater rate of oxygen consumption than the intact seed, and that the germination of the upper seed was stimulated in atmospheres enriched in oxygen. However, Porter and Wareing discovered that oxygen can diffuse through the seed coats of both upper and lower seeds faster than it can be consumed by the embryos, and that upper and lower seeds are closely similar with respect to their oxygen consumption and the oxygen permeability of their seed coats. Previously, Wareing and Foda (1957) had reported that upper seed embryos remained dormant when provided with moisture from a saturated atmosphere, but germinated on moist filter papers as a result of the elution of germination inhibitors by the water held on the filter paper. Porter and Wareing concluded that the upper seeds are not dormant because of the seed coat restriction of oxygen diffusion.

In the literature I am not able to find convincing evidence that the restriction of gaseous exchange by the embryo coverings brings about dormancy.

The restriction of water uptake by embryo coverings

Hard seeds (which occur in a number of families, although they are most common in the small-seeded Leguminosae) are those seeds which fail to imbibe water during the two or three weeks normally considered sufficient for the duration of a germination test. An impermeable seed coat favours the maintenance of a constant moisture content in the embryo, which if optimal for seed survival is clearly advantageous (see pp. 106–107). Hyde (1954) investigated the simultaneous occurrence of two apparently incompatible processes, the development of an impermeable testa and the desiccation of the embryo. In three species from the Leguminosae (*Trifolium repens*, white clover, *T. pratense*, red clover, *Lupinus arboreus*, tree lupin) Hyde found that the fissure in the hilum acted as a hygroscopic valve by remaining closed at high relative humidity but opening at low relative humidity, so that the seed was permitted to dry out. In the tree lupin seeds, Hyde observed that the hilar fissure opened or closed within one minute of the appropriate change in the relative humidity. The open fissure showed a gap of between 15 and 75 μm. Eventually the hard seeds tended to attain a moisture content in equilibrium with the lowest relative humidity to which they had been exposed. Such seeds absorbed moisture under conditions of gradually increasing relative humidity such that the fissure remained open. As such a course of events is likely to be infrequent in the soils of arid regions, leguminous seeds may remain dormant in the soil for years.

As the imbibition of seeds results in a substantial increase in embryo dimensions, there is normally no difficulty in recognizing the occurrence of water uptake. However, in some hard seeds it is difficult to distinguish between the restriction of water uptake and the mechanical restriction of embryo expansion (see below) as the cause of dormancy. One way of studying the diffusion of water into seeds is to follow the uptake of ^3H_2O by means of autoradiography, as described by Jackson and Varriano-Marston (1980).

Mechanical restriction of embryo growth

The embryo coverings frequently provide a mechanical restriction to embryo growth. Crocker and Davis (1914) showed that the seed coat of *Alisma plantago* prevented the complete imbibition of the embryo so that the seed with a partly swollen embryo could lie for years in its aquatic environment, unable to germinate. These workers did not explain how mechanical restriction may eventually be overcome in *Alisma*, although one may infer that natural events resulting in physical damage to the restricting layers may be effective (see pp. 71–72). In other species it has been shown that development of the embryo overcomes the mechanical restriction of the embryo coverings, thus implying that there are two aspects of this form of dormancy, the mechanical restriction of the coverings and the physiological ability of the embryo to overcome this restriction. For the dormant upper seed of *Xanthium*, Esashi and Leopold (1968) showed that the radicle was unable to develop sufficient thrust to cause testa rupture. For lettuce, Ikuma and Thimann (1963) demonstrated that the mechanical properties of the endosperm layer prevented radicle elongation in seeds which had not received an irradiation treatment sufficient to bring about germination. They also obtained some supporting evidence for their hypothesis that the tip of the radicle becomes able to penetrate the endosperm as a consequence of the secretion of a hydrolysing enzyme. For *Eucalyptus pauciflora* and *E. delegatensis* seeds, in which he deduced that the primary cause of dormancy was the mechanical resistance of the seed coat, Bachelard (1967) pointed out that embryos excised after long periods of imbibition showed no evidence of radicle growth or distortion. This observation seems to be generally true of seeds showing this form of dormancy, as is Bachelard's conclusion 'that the germination processes can be initiated and can develop to a certain stage in imbibed intact seeds but when resistance to germination is encountered, in this case by a mechanically resistant seed coat, further stages in the germination process are prevented'.

Not all mechanically resistant embryo coverings are subjected to penetration by the radicle. In hazel nuts, the pericarp is split at the margins of the two carpels constituting the ovary by the expansion of the embryo, mainly the cotyledons (see p. 57).

Water-soluble inhibitors in the embryo coverings

Although there are very many examples in which the growth and development of whole plants or their isolated cells or organs have been shown to be inhibited by plant constituents, there are comparatively few examples where the physiological role of an endogenous inhibitor has been established firmly. If a germination inhibitor functions at concentrations which are micromolar or less, and in addition if it shows its physiological effect in a cell, tissue or organ other than that in which it was synthesized, the inhibitor should be regarded as a plant growth substance. Traditional hypotheses about the mode of action of plant growth substances have assumed that growth is regulated by the concentrations *in vivo* of the growth substances. However, as evidence consistent with traditional hypotheses is rarely found, support has grown for the concept that the physiological effects of plant growth substances may result from changes in cellular sensitivity to them (Trewavas, 1982). Therefore the criteria for the role of endogenous inhibitors in seed dormancy may be stated thus:

(i) There should be a correlation between the state of dormancy and either the concentration of endogenous inhibitor or the sensitivity of the embryo to the inhibitor
(ii) Exogenous application of inhibitors should inhibit germination reversibly, although changes in the sensitivity of the embryo to the inhibitor may also be related to the depth of dormancy.

These criteria represent both a simplification and an extension of the PESIGS rules formulated by Jacobs (see Jacobs, 1962) to determine adequate evidence as to the internal factor (plant growth substance) which normally controls a given process.

Since the effect of germination inhibitors must occur within the embryo, research workers directed their attention to the inhibitor content of the embryo. However, as traditional procedures for the extraction and identification of plant growth substances demand large quantities of starting materials (from one to many kg), most workers extracted inhibitors from whole seeds or fruits, so that in only a few instances was the inhibitor shown to be located in the seed coverings, as shown in Table 5.4, or in the embryo, as shown in Tables 5.4 and 5.5 (see Wareing, 1965). Another flaw in much of the published work is that suspected inhibitors have not usually been shown to inhibit the germination of the species from which they have extracted, indeed they have frequently not been tested on the germination of any species, having been tested only by one or more of the standard bioassays such as wheat embryo or wheat coleoptile growth.

As an example I shall consider the role of inhibitors in the dormancy of hazel seeds. Bradbeer (1968) inferred that the papery testa contained water-soluble inhibitors which prevented the germination of the non-dormant newly-harvested embryo. These inhibitors were presumed to be transferred from the dead cells of the testa to the embryo during imbibition. Evidence in support of this hypothesis is shown in Table 5.3. Laceration of the testa was ineffective in promoting germination, while complete removal of the testa permitted substantial germination. The presence of water-soluble inhibitors of germination was demonstrated by placing the detached testas

Table 5.3 The effects of hazel testa and pericarp on germination

Treatment	% germination of newly harvested seeds (non-dormant embryo) after 8 days at 20 °C[1]	% germination of dormant seeds (dormant embryo) after 12 days at 20 °C[2]
Intact seed	5	0
Intact seed with lacerated testa	5	—
Embryo (testa removed from petri dish)	51	—
Embryo (testa in petri dish)	33	—
Embryo (testa + pericarp in petri dish)	25	—
Embryo in $3 \cdot 10^{-5}$ M ABA	15	—
Intact seed in 10^{-4} M GA_3	—	45
Intact seed in 10^{-4} M GA_3 plus 1 extra testa per seed	—	25
” ” ” ” ” ” 2 ” ”	—	15
” ” ” ” ” ” 3 ” ”	—	7

[1] From Bradbeer (1968)
[2] From Ross, J.D., unpubl. Ph.D. thesis, University of London (1971)

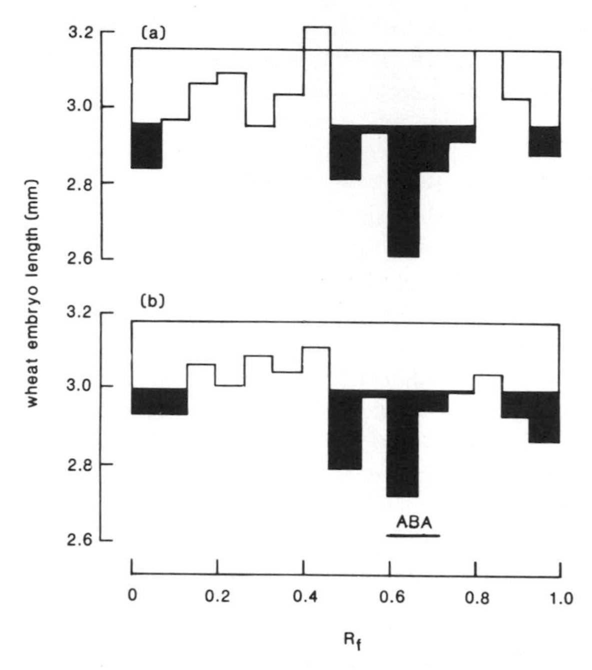

Figure 5.1 Histograms representing the wheat embryo bioassay of the thin layer chromatograms of the acidic diethyl ether fractions obtained from the pericarp and testa of hazel. Darkened areas indicate growth significantly different ($p = 0.05$) from control. The position of ABA on each set of chromatograms is marked. *a*, pericarp from 6 freshly harvested nuts; *b*, testa from 10 freshly harvested nuts. After Williams *et al.* (1973).

and pericarps with the embryos in the petri dishes and observing their inhibitory effects. Furthermore, J.D. Ross showed that when extra testas were added to petri dishes in which dormant hazel seeds were caused to germinate in 10^{-4}M GA_3, the presence of the testas brought about inhibition of germination.

As both aqueous and methanolic extracts of hazel testa and pericarp inhibited hazel germination, an attempt was made to characterize the inhibitors. For convenience the wheat embryo bioassay was used for the detection of inhibitors. Thin layer chromatograms of the acidic diethyl ether fractions obtained from methanolic extracts of hazel testas and pericarps each showed four zones of inhibitory activity (Figure 5.1) (Williams *et al.*, 1973). One of the inhibitory zones co-chromatographed with ABA, and the identification was confirmed by GLC and GLC/MS, which also showed that the ABA content of the extracts could be accurately determined by GLC at the level of accuracy normally associated with physico-chemical methods. In contrast, growth substance bioassays are very much less reliable, as they are subject to many possible sources of error. From Figure 5.1 it can be seen that the testa and pericarp contained similar amounts of inhibitors, but as this batch of nuts had their dry weights distributed as 1600 mg of pericarp, 80 mg of testa and 1100 mg of embryo, it is clear that the greatest concentration of inhibitors was in the testa.

The effect of exogenous ABA on the germination of freshly harvested hazel embryos (Table 5.3) is consistent with a germination-preventing role for the ABA in the embryo coverings. Williams and co-workers (1973) demonstrated falls in the ABA contents of hazel embryo covering layers amounting to about 9 pmoles/day/nut during dry storage and about 40 pmoles/day/nut during imbibition, which are consistent with the postulated role of ABA in hazel seed dormancy. Embryo dormancy develops in hazel as the moisture content falls during ripening, after which stage ABA does not seem to be necessary for the maintenance of dormancy. However, further work is required to determine the mechanism by which ABA prevents germination and also to explore the role of the other inhibitors of the hazel seed coverings. The inhibition of wheat embryo growth by these substances does not prove that they are involved in hazel seed dormancy.

Table 5.4 is a summary of the best available examples in which identified water-soluble inhibitors have been shown to occur in the embryo coverings of dormant seeds. It is important to note that the unequivocal identification of an endogenous inhibitor requires the use of physico-chemical methods. ABA is involved in five of these seven species, in four of which exogenously supplied ABA has been shown to inhibit the germination of the non-dormant seeds of the same species. Furthermore, the reported ABA concentration in the coverings of these seeds appears to be high enough to maintain embryo dormancy. Indeed, the concentrations of ABA listed in Table 5.4 are among the highest naturally occurring concentrations known. The available evidence indicates that ABA is able to prevent the commencement of germination and that this inhibition is reversible. Reversibility may result, for example, from the enzymic inactivation of ABA, from leaching, or from a change in the sensitivity of the embryo to ABA, or from the formation of germination promoters.

Table 5.4 lists two species containing phenolic acids and one species containing a coumarin-like substance. The phenolic acids of sugar maple are present in concentrations up to two orders of magnitude greater than that of ABA. Neither the coumarin-like substance nor any of the phenolic

Table 5.4 Water-soluble germination inhibitors shown to occur in embryo coverings and embryos of dormant seeds

Species	Substance	Location and concentration of inhibitors (nmoles/g. dry weight) in fruit and seed	Source
Malus sylvestris (apple)	ABA[1,3]	Testa	Pieniazeck and Grochowska (1967)
Corylus avellana (hazel)	ABA[2,3]	Pericarp 1.4 Testa 19	Bradbeer (1968)
		Embryo 0.09	Williams *et al.* (1973)
Rosa arvensis (field rose)	ABA[2]	Flesh of rose hip 15.5[5] Mature achene 0.6[5]	Jackson (1968)
Fraxinus americana (white ash)	ABA[2,3]	Pericarp from dormant samara 2.8 Dormant seed 1.7	Sondheimer *et al.* (1968)
Elaeagnus umbellata (autumn olive)	Coumarin-like substance[1,4]	Apparently present in similar concentrations in exocarp, endocarp testa and embryo	Hamilton and Carpenter (1975)
Comptonia peregrina (sweet fern)	ABA[1,3]	In testa	Del Tredici and Torrey (1976)
Acer saccharum (sugar maple)	ABA[2,3]	Pericarp, 0.83; testa, 5.1; cotyledons, 1.9; axis,2.8	Enu-Kwesi and Dumbroff (1978)
" "	Ferulate[1,4]	Pericarp, 67; testa, 124; cotyledons, 21; axis, 216.	Enu-Kwesi and Dumbroff (1980)
" "	*o*-Coumarate[1,4]	Pericarp, 55; testa, 329; cotyledons, 43; axis, 348.	"
" "	*p*-Coumarate[1,4]	Pericarp, 189; testa, 713; cotyledons, 244; axis, 799.	"
Parthenium argentatum (guayule)	*p*-Hydroxybenzoate[1] Protocatechuate[1] *p*-Coumarate[1] Ferulate[1] Benzoate[1] Vanillate[1] Cinnamate[1]	In the chaff	Naqvi and Hanson (1983)

[1] Provisional identification.
[2] Unequivocal identification by physico-chemical methods.
[3] Inhibits germination of non-dormant seeds of this species.
[4] Does not inhibit germination of non-dormant seeds of this species.
[5] nmoles/g fresh weight.

acids have been reported to inhibit germination when supplied exogenously to the seeds of the same species, and it remains to be established whether they are reversible inhibitors of germination as would be necessary if they play a role in seed dormancy. Such substances may protect seeds against micro-organisms or have an allelopathic function against seeds or roots of other species.

It is interesting to note that all of these species listed in Table 5.4 are trees or shrubs, although it may be that woody plants are more likely to have large seeds which permit the easy separation of the embryo from the covering structures.

Germination inhibitors have been shown to occur in embryos (see pp. 50–52), in which case the embryo coverings may maintain dormancy by the prevention of the release of germination inhibitors. Wareing and Foda (1957) reported that the semipermeable testa of *Xanthium pennsylvanicum* prevented the leaching of germination inhibitors from the embryo. Webb and Wareing (1972*a*) showed that if the testa of *Acer pseudoplatanus* was torn and if the seed was placed on moist filter-paper so that the hole was in contact with liquid water, leaching of inhibitor occurred and germination was rapid.

Dormancy from the failure to mobilize extra-embryonic food reserves

In dormant seeds whose major food reserves are extra-embryonic (in endosperm or perisperm, see pp. 18–22) detectable mobilization of these reserves may be reduced or absent. In *Heracleum sphondylium* there was no germination at 15 °C because hydrolysis of reserve proteins in the endosperm did not occur at this temperature, although there was some hydrolysis of reserve polysaccharides (Stokes, 1953*a*). Stokes (1953*b*) showed that isolated embryos germinated and grew normally in culture in a mineral medium containing 2% glucose and an appropriate nitrogen source (KNO_3 or a suitable amino acid). Normal germination of the dormant seed, including the hydrolysis of endosperm proteins, occurred in seeds chilled at 2 °C. Stokes (1953*a*) deduced that some product of embryo metabolism at 2 °C diffused into the endosperm and brought about protein hydrolysis. This situation is analogous to the germination of graminaceous seeds in which gibberellins diffuse from the embryo to bring about the synthesis, activation and secretion of hydrolytic enzymes by the aleurone cells, followed by the hydrolysis of reserve polysaccharides, proteins, lipids, etc., and the absorption of the soluble products by the scutellum of the germinating embryo. Thus failure to germinate may be regarded as a block to the mobilization of extra-embryonic reserves, although the cause of this may be a failure of the embryo to secrete an activating factor.

Embryo dormancy

Isolated embryos which are capable of germination and growth when supplied with sufficient moisture and a suitable constant temperature are clearly not dormant. Embryos which require basic nutrients (minerals and a carbon and nitrogen source) for germination, as for *Heracleum sphondylium* discussed above, are also not dormant. However, as the embryo may not, at this stage, be capable of mobilizing the extra-embryonic food reserves, the cause of the seed dormancy may lie within the embryo. At the present time the mechanisms of embryo dormancy are not well defined and no example has been elucidated in molecular terms. Mechanisms or perhaps groups of mechanisms have been proposed (Table 5.2) and are worthy of consideration.

Underdeveloped and undifferentiated embryos

Many authors have considered 'embryo immaturity' to be a cause of dormancy, although the present writer suggests that the implication that embryos can be mature is a semantic contradiction. What is meant is that many seeds contain a relatively small embryo at the time of dehiscence, so that a considerable period elapses between imbibition and radicle protrusion. As an example, *Heracleum sphondylium* seeds at dehiscence possess very small heart-shaped embryos comprising 0·4% of the dry weight of the seed, which before germination grow to embryos weighing 30% of the total seed weight (Stokes, 1952). Although the initial stages of embryo growth occurred either at 2 °C or 15 °C, germination occurred at 2 °C and not at 15 °C. A somewhat different situation is shown by some species of *Fraxinus*, e.g. *F. nigra* (Steinbauer, 1937) and *F. excelsior* (Villiers and Wareing, 1964), which contain fully differentiated embryos of $\frac{1}{3}$ to $\frac{1}{2}$ of the length reached prior to germination. Embryo growth within the seed is favoured by a temperature of 20–30 °C but a 2–3 month chilling period at 5–10 °C is subsequently required before germination will occur. Such mechanisms increase the time between dehiscence and germination, as for example with some *Fraxinus* species in which germination is delayed until the second spring after dehiscence.

The mechanism of embryo dormancy

Since radicle elongation is the usual criterion for recording the occurrence of germination, it is necessary to consider whether the mechanism of embryo dormancy involves a direct inhibition of radicle growth. This question can be explored by attempting to grow either the whole embryonic axis (radicle and plumule) or just the radicle in culture. Research on plant cell and organ culture has shown that the selection of defined culture media, normally containing plant growth substances, is usually critical. An optimistic opinion would be that it should be possible to grow in culture any plant cell or organ which is capable of growth and development *in vivo*. Although a great deal of work has been carried out on the culture of ovules and of embryos at all stages of development (Raghavan, 1976), comparatively little has been done with embryonic axes from dormant embryos other than hazel.

Jarvis *et al.* (1978) obtained considerable embryonic axis growth and development when axes from dormant hazel embryos were cultured on the inorganic medium of Tukey supplemented with 2% sucrose. Subsequent work has shown that similarly treated embryonic axes can give rise to seedlings which can be grown-on autotrophically (Jarvis, pers. comm.). As negligible growth occurred on inorganic media, it must be concluded that the radicles and plumules of dormant hazel embryos are not intrinsically dormant and the immediate cause of hazel embryo dormancy must be an inability to mobilize the reserve materials of the embryo. Our earlier work showed that the imbibed but dormant hazel embryo (about 60% moisture content) showed no detectable mobilization of reserves, although the respiration and the metabolism of radioactive substrates by both axes and cotyledons was substantial (Bradbeer and Colman, 1967; Bradbeer and Pinfield, 1967). Hazel embryo dormancy can be broken by exogenous GA_3 or by a treatment which results in the synthesis of endogenous GA (6 weeks's chilling at 5 °C followed by transfer to higher temperature such as 20 °C) (see pp. 56–64). When dormancy is broken in these ways, the embryonic reserves are mobilized. Hence in dormant hazel embroys, dormancy may be considered to result from a lack of GAs which may in turn result from an insufficiency of GA synthesis or an inability to release bound GA, or rapid consumption of free GA in the embryo. For hazel, there is evidence that the embryonic axis is the main site of gibberellin synthesis and that there is some synthesis and some release of bound gibberellins in the cotyledons (Arias *et al.*, 1976).

Beyond this level a molecular explanation of seed dormancy is lacking and we can do little more than indulge in speculation. Developmental changes in plants, such as photomorphogenesis and chloroplast development, have been shown to involve changes in protein synthesis at the levels of genome transcription, the translation of messenger RNA and the post-translational modification of the newly synthesized polypeptides (see for example, Bradbeer, 1981). There are a number of cases in which the breaking of embryo dormancy has been shown to involve the synthesis of one or more new polypeptides by the embryo, together with the appearance of the corresponding mRNAs. However, it has not been shown whether such changes in protein synthesis are responsible for the breaking of embryo dormancy or are a consequence of the breaking of dormancy. Furthermore, such investigations of protein synthesis and mRNA have so far concerned only the 20–50 more abundant molecules out of several thousand polypeptides coded by the genome of each plant. Even those polypeptides which have been detected, by means of polyacrylamide gel electrophoresis, in dormant embryos, have not been identified in terms of biological function. It is possible that the immediate cause of embryo dormancy does not lie in the deficiency of a protein, but for example in the regulation of enzyme activity or in membrane permeability or membrane transport. Further speculation seems inappropriate here.

Endogenous inhibitors in embryo dormancy

Although there has been considerable interest in the role of germination inhibitors in seed dormancy there are very few firmly established cases of the presence, in dormant embryos, of unequivocally identified substances which are able to inhibit the germination of the non-dormant seeds of the same species. In Table 5.5 there is one example, namely ABA in apple embryo, and two more cases are included in Table 5.4: ABA in the embryos of hazel and sugar maple. Table 5.5 also lists the next best examples of this kind, notably the unidentified germination inhibitors in the embryos of ash and sycamore, and ABA clearly detected in yew embryos but not reported to inhibit the germination of the dormant seed. Provisional identifications of *trans-p*-coumaric acid in embryos of spindle tree and of ABA in lettuce and pine are accompanied by proof of ability to inhibit germination only in the case of lettuce. The use of physico-chemical methods to identify suspected inhibitors also permits accurate measurement of inhibitor concentrations. The three species for which embryo concentrations of ABA can be determined from Tables 5.4 and 5.5 fall approximately in the range of 3×10^{-6}M for sugar maple, 3×10^{-7}M for apple and 1×10^{-7}M for hazel. This is based on the assumption that the ABA is evenly distributed through the embryo, which is clearly not the case for free ABA of apple embryo. ABA in the embryo coverings is present in rather higher amounts (see Table 5.4 for ABA expressed in nmoles/g dry weight) as would be expected if the ABA was to dissolve in the water involved in the imbibition of the embryo.

In the cases of yew and apple, ABA was shown to be present in both free and bound forms, and it was demonstrated that the bound form could be

Table 5.5 Germination inhibitors in embryos of dormant seeds. See Table 5.4 for some other examples.

Species	Substance	Location and concentrations of inhibitors (nmoles/g fresh weight)	Source
Xanthium pennsylvanicum (cocklebur)	Two unidentified inhibitors	Embryo	Wareing and Foda (1957)
Fraxinus excelsior (ash)	Unidentified[1,2]	Embryo and endosperm	Villiers and Wareing (1965)
Euonymus europaeus (spindle tree)	*trans-p*-coumaric acid[3]	Embryo	Monin (1967)
Acer pseudoplatanus (sycamore)	Unidentified neutral compound[1]	Embryo	Webb and Wareing (1972*b*)
Taxus baccata (yew)	free ABA[4] bound ABA[4]	Embryo (approximately equal amounts of free and bound forms)	Le Page-Degivry (1973)
Lactuca sativa (lettuce)	ABA[1,3]	Embryo	Speer and Tupper (1975)
Malus sylvestris (apple)	Free ABA[1,4]	Axis, trace; Cotyledons, 0.34.	Le Page-Degivry and Bulard (1979)
	Bound ABA[1,4]	Axis, 0.26; cotyledons 0.18	
Pinus pinea (stone pine)	ABA[3]	Embryo and endosperm	Corvillon and Martinez-Honduvilla (1980)

[1] Inhibits germination of non-dormant seeds of this species.
[2] Found only in imbibed seed.
[3] Provisional identification.
[4] Unequivocal identication by physico-chemical methods.

hydrolysed *in vivo* to give the free form. There seems to be a reasonably strong case that ABA can maintain an embryo in a state of dormancy (see p. 45 for further discussion). Williams *et al.* (1973) found that the ABA content of dormant hazel embryos was not significantly different from that of freshly harvested non-dormant embryos, and that the ABA content of the embryos fell under dry or moist conditions, irrespective of whether embryo dormancy was broken by the treatment given. Thus in hazel, ABA seems to maintain freshly-harvested non-dormant embryos in a dormant condition, but subsequently the sensitivity of the embryos to ABA falls and ABA ceases to regulate dormancy. In sugar maple the ABA content of embryo and embryo coverings fell during chilling at 5 °C, results which were consistent with a role for ABA in seed dormancy (Enu-Kwesi and Dumbroff, 1978). In contrast, Villiers and Wareing (1965) found that the chilling treatment, which was necessary for the breaking of ash seed dormancy, did not remove inhibitor from either the seed or the embryo. If the inhibitor in ash seed is partly responsible for ash seed dormancy, then the breaking of dormancy presumably involves a change in sensitivity to inhibitors. Firm evidence that identified seed or fruit constituents other than ABA are definitely involved in seed dormancy is not available.

The possession of more than one dormancy mechanism

Among those seeds or fruits whose dormancy has been investigated more intensively, a number have been found to possess more than one of the dormancy mechanisms listed in Table 5.2. Clearly distinct mechanisms can be seen, and some examples are included in Table 5.6. For inclusion in this category it is not sufficient to possess more than one way of breaking dormancy, as a single type of dormancy may be broken in several different ways. Inadequate knowledge of the biochemical mechanisms of dormancy has so far prevented any demonstration that there may be more than one distinct biochemical block in the embryo. One might speculate that the embryo mechanisms listed in Table 5.2 may be linked, as a deficiency of nucleic acid or protein synthesis, for example, may result from a failure to mobilize food reserves which may in turn have resulted from a lack of plant growth substances or the presence of inhibitors.

In hazel, the possession of three dormancy mechanisms seems to provide a more precise control of germination. During the development of any seed on its parent plant, the developing embryo is in a state of rapid growth, so that if released from the parent plant it is common for the embryo to continue into a state of premature germination. In hazel, the mechanical restrictions imposed by the pericarp and the presence of inhibitory material

Table 5.6 Some examples of seeds and fruits which possess more than one dormancy mechanism

Species	Dispersal unit	Dormancy mechanisms	Source
Xanthium pennsylvanicum (cocklebur)	Seed	Embryo dormancy (endogenous inhibitors). Testa prevents leaching of inhibitors. Mechanical restriction by testa.	Wareing and Foda (1957) Esashi and Leopold (1969)
Fraxinus excelsior (ash)	Samara (fruit)	Underdeveloped embryo. Pericarp impermeable to O_2. Inhibitor accumulates in embryo during imbibition.	Villiers and Wareing (1964)
Elaeagnus umbellata (autumn olive)	Drupe (fruit)	Mechanical restriction by pericarp. Inhibitor in pericarp, testa and/or embryo. Embryo dormancy.	Hamilton and Carpenter (1975)
Avena fatua (wild oat)	Caryopsis (fruit)	Inhibitor in hull (lemma and palea) enclosing caryopsis. Embryo dormancy. Hulls, pericarp and testa may interfere with water uptake and gas exchange.	Simpson (1978)
Corylus avellana (hazel)	Nut (fruit)	Mechanical restriction by pericarp. Inhibitor in embryo coverings. Embryo dormancy develops as seed dries during ripening.	Bradbeer *et al.* (1978)

in the seed coverings provide a sufficient guarantee of the maintenance of dormancy. As embryo dormancy develops in the drying seed, the requirement for control by inhibitors is reduced and the sensitivity of the embryo to inhibitors falls. The possession of a hard pericarp delays germination until the chilled cotyledons are able to exert enough pressure to rupture this outer covering. Embryo dormancy and the mechanical restriction to embryo growth are broken by a sufficiently long chilling treatment, after which germination can be accelerated by a raised temperature (20 °C).

The number of entries possible in Table 5.6 is presumed to be limited by the small number of seeds and fruits on which intensive studies have been made.

CHAPTER SIX

THE BREAKING OF SEED DORMANCY

Dormancy may be broken as a result of the exposure of the seed to a single factor at the requisite intensity for an appropriate period of time. Germination then follows the breaking of dormancy. Alternatively, dormancy breaking may require or may be accelerated by exposure of the seed to fluctuating conditions, such as diurnal changes of soil temperature or the diurnal cycle of light and darkness. Dormancy may be broken in response to the range of diurnal fluctuations occurring at a particular season of the year, such as spring, or dormancy breaking may result from the seasonal cycle of temperature, as for example in the succession through winter, summer and winter. In a more complex situation, dormancy may be broken as a result of exposure of seeds to two or more distinct factors, such as light and an appropriate temperature range, or to ethylene at an appropriate temperature. Table 6.1 lists the main types of dormancy-breaking treatments.

Dormancy breaking by single factors

In my experience, the majority of dormant seeds requires the application of more than one factor before dormancy can be broken. The investigation of the mechanism of dormancy breaking may be more straightforward in seeds where the application of a single factor is sufficient.

Table 6.1 The main ways in which seed dormancy may be broken.

A. Exposure to a constant single factor
  1. Chilling
  2. Dry storage/elevated temperatures
  3. Light
  4. Leaching
  5. Scarification
  6. Exposure to chemicals

B. Exposure to fluctuating conditions

C. Exposure to two or more factors

Chilling

Chilling involves the exposure of imbibed seeds to low temperatures, normally between 1° and 0 °C, for what may be a substantial period. It is carried out in forestry nurseries by overwintering seeds or fruits in shallow pits, a process frequently described as 'stratification', because the seed material is placed in alternate layers with layers of sand or soil. The seeds are protected from predation with wire mesh and are dug up in early spring for sowing in seed beds. Figure 6.1 shows the results of a stratification experiment carried out with hazel nuts during a winter in which the soil was frozen on occasions. When seeds were used in the germination tests, 12 weeks' stratification was sufficient for fairly complete germination, but when nuts were germinated, a rather longer stratification period was required. The greater chilling requirement of the intact nut probably results from its possession of three dormancy mechanisms: dormant embryo, hard pericarp and inhibitors in testa and pericarp. The intact seed has only embryo dormancy and the inhibitors in the testa as dormancy mechanisms (see following paragraph, and Chapter 5).

Horticulturalists achieve the same results as those obtained by stratification by overwintering seeds in seed pans and looking for germination in spring or summer. Better control is obtained experimentally by carrying

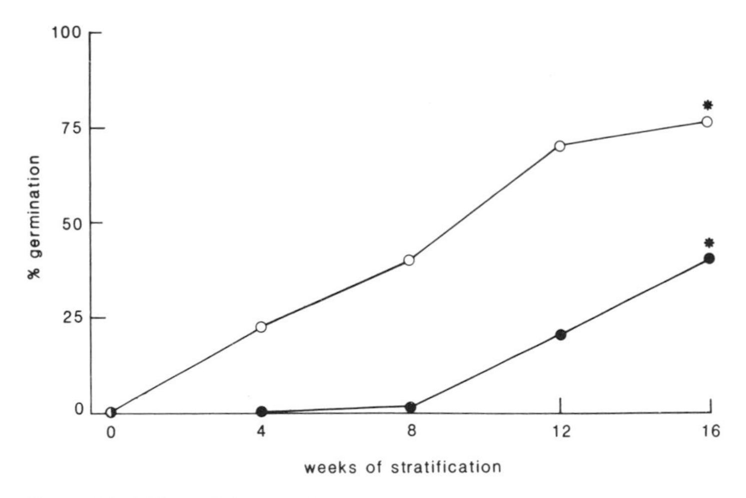

Figure 6.1 Effect of the stratification of hazel nuts (Kent cobnuts) on the subsequent germination of seeds (○) and nuts (●) during 28 days' incubation at 20 °C. Stratification commenced in late November 1979, in London, with the nuts under 10 cm of soil. *Includes some germination which took place in the soil. Results of E.M. McMorrow.

out the chilling in a refrigerator or cold room, larger seeds being held in seed trays between layers of sand or vermiculite and smaller seeds being mixed with moist sand in closed specimen tubes or jars.

In our routine laboratory investigations, a refrigerated incubator set at 5 °C provides our standard chilling treatment, and germination subsequent to chilling is determined by transfer of the seeds to 20 °C. In hazel nuts, seed dormancy is normally broken by chilling the intact nuts in boxes of moist vermiculite at 5 °C for 4–6 weeks, after which rapid germination occurs if the seeds are removed from the pericarp and incubated at 20 °C (see Figure 6.2*a*). As is also shown in Figure 6.1, when germination tests are conducted with intact nuts a longer chilling treatment is required (see Figure 6.2*b*), because in this case germination also involves the rupture of the pericarp in response to expansion of the cotyledons and there is the possibility that inhibitors arising from the pericarp may increase the depth of dormancy. The seeds used in the experiment illustrated in Figure 6.2 were not quite fully dormant, in that 7% germination occurred in the absence of any chilling treatment. With increased duration of chilling, both the rate of germination and the total germination increased. Many seeds, including hazel, will germinate at the chilling temperature after a rather prolonged period. Although the data in Figure 6.2 are typical for Kent cobnuts, a considerable range of variation is found with respect to the depth of seed dormancy: this changes during ripening and storage of the seeds, varies from season to season, and is dependent on seed provenance. Such variation in depth of dormancy is a property typical of seeds

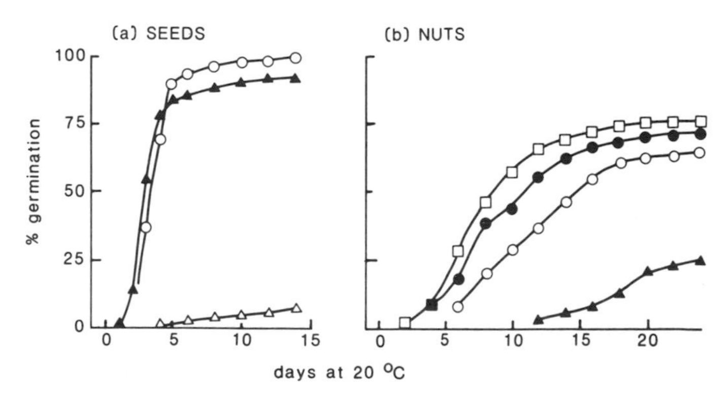

Figure 6.2 Germination of hazel seeds (*a*) and nuts (*b*) at 20 °C after various periods of chilling (as intact nuts) at 5 °C in moist vermiculite. Chilling for 6 weeks, ▲; 9 weeks, ○; 12 weeks, ●; 18 weeks, □. No chilling, △. 1980 crop, after G.A. Rendon (unpubl. Ph.D. thesis, University of London, 1983).

irrespective of the mechanisms by which their dormancy is broken. Genetic differences in seed dormancy seem to have been recorded within all species in which the dormancy of different genotypes has been compared, as for example in the intensive studies on *Avena fatua* (Simpson, 1978).

It should be noted that the post-chilling rise of temperature necessary to bring about a fairly rapid germination of chilled hazel seeds could be regarded strictly as a result of a fluctuation of environmental conditions. However, since chilling is the critical factor, its consideration will remain in this section.

The effects of chilling on a number of species have been considered by several reviewers (e.g. Crocker and Barton, 1953; Stokes, 1965; Nikolaeva, 1967, 1977; and Bewley and Black, 1982). Here we shall consider in detail the mechanism of dormancy-breaking in response to chilling in one species, hazel.

The mechanism of dormancy breaking during chilling of hazel seeds

In 1958, B. Colman and the author entered into co-operation with P.F. Wareing and his then postgraduate student, B. Frankland, with the objective of elucidating the mechanism by which chilling brought about the breaking of seed dormancy. We needed a fairly large seed to facilitate biochemical studies and preferably a seed which could be dissected into its separate parts. Our first choice was beech (*Fagus sylvatica*) with which Colman set up an initial set of metabolic studies. Frankland was already involved in an investigation of the effect of chilling on the endogenous gibberellins of beech seed. A complete failure of beech seed production in 1959 forced us all to choose hazel seed as an alternative material which was expected never to suffer a total crop failure.

Bradbeer and Colman investigated the metabolism in hazel seed of [2-^{14}C]-acetate at 5 °C, using cotyledon slices and isolated embryonic axes obtained at various stages of the chilling process. In dormant hazel seeds the major pathways of intermediary metabolism were functional, and there was no evidence for the presence of any metabolic blocks. As chilling progressed, the labelling of proteins and nucleic acids increased, but, as nucleic acid and protein syntheses are generally agreed to be early processes in germination, we were unable to identify metabolic events which might have been part of the dormancy-breaking process (see p. 49, and Bradbeer and Colman, 1967). Further investigations on the metabolic consequences of the provision of GA_3 to dormant hazel seeds (Bradbeer and Pinfield, 1967) and on the nucleic acids of the seeds (Wood and Bradbeer, 1967) did not lead to a decisive solution.

In his parallel investigations on GA-like substances, Frankland had encountered considerable technical difficulties in the bioassay of the very small amounts of GA-like activity that he extracted from hazel seeds, especially as a result of the presence of substances in the extracts which proved phytotoxic in the bioassays. The main findings of Frankland and Wareing (1966) were that dormant hazel seeds contained no detectable GA-like substances, but that after 12 weeks' chilling the hazel seeds contained the biological equivalent of 0.2 pmoles of GA_3. They suggested that GA synthesized during chilling was responsible for the breakage of hazel seed dormancy, even though the concentrations of endogenous GA-like material were much lower than the concentrations of exogenous GA_3 which were capable of bringing about the germination of the dormant seeds. Bradbeer and Pinfield (1967) had found that the application of 10 nmoles of GA_3 at a concentration of 2.5×10^{-6} M induced 30% of dormant hazel seeds to germinate. The discrepancy between the amount of exogenous GA_3 required to break hazel seed dormancy and the 0.2 pmoles of endogenous GA-like material detected by Frankland amounted to five orders of magnitude, as shown in Figure 6.8. Bradbeer (1968) therefore tested the hypothesis of Frankland and Wareing (1966) by chilling dormant hazel seeds for 28 days in 6×10^{-4} M chlorocholine chloride (CCC), an inhibitor of GA synthesis, germination being tested subsequently at 20 °C in water: Table 6.2 shows that CCC was without effect in this experiment, but that it significantly inhibited the germination of normally chilled hazel seeds when supplied at 20 °C during the germination test. It was concluded

Table 6.2 The effects of the time of application of CCC (chlorocholine chloride) on the breaking of hazel seed dormancy by 28 days of chilling at 5 °C followed by 15 days at 20 °C. (1), no CCC; (2), with 10^{-6}M CCC. Data from Bradbeer (1968).

Chilling treatment				
Material chilled	Seeds		Intact nuts	
Chilling medium	Water		Moist sand	
Germination treatment				
Material treated	Seeds		Seeds	
Germination medium	Water		Water	
	% Germination			
	(1)	(2)	(1)	(2)
5 days at 20 °C	26	23	45	23
10 days at 20 °C	47	49	64	50
15 days at 20 °C	50	52	68	51

In this experiment the 95% confidence limits were ± 14%.

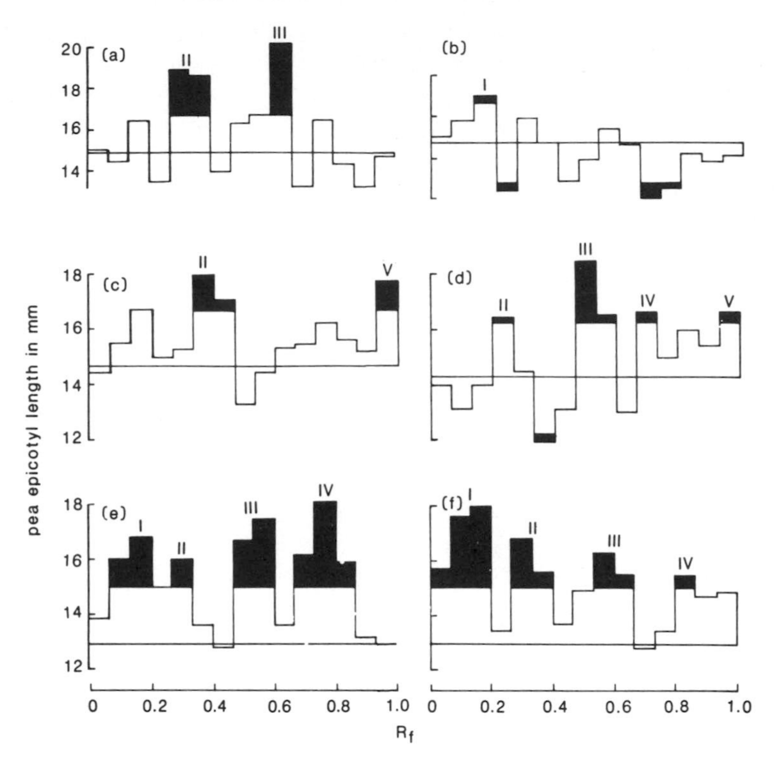

Figure 6.3 Histograms representing the GA bioassay with excised epicotyls of dwarf peas of the thin layer chromatograms of the acidic diethyl ether fractions obtained from hazel embryos of seeds which had successively developed dormancy, been chilled at 5 °C and transferred to 20 °C. Darkened areas indicate growth significantly different ($p = 0.05$) from control (bioassay of blank chromatogram). TLC on 0.3 mm layer of Merck Kiesel P240 developed in ethyl acetate: chloroform: acetic acid (15:5:1). (*a*) 15 freshly harvested embryos (non-dormant); (*b*) 2 embryos after 28 days dry storage (dormant); (*c*) cotyledons of 8 chilled seeds; (*d*) embryonic axes of 25 chilled seeds; (*e*) cotyledons of 5 chilled seeds, extracted after 7 days at 20 °C; and (*f*) embryonic axes of 20 chilled seeds extracted after 2 days at 20 °C. After Ross and Bradbeer (1971*a*).

that if GA synthesis has an important role in the breaking of hazel seed dormancy, then synthesis must occur subsequently to chilling when the temperature is raised to a level favourable for germination.

Consequently, J.D. Ross commenced a re-investigation of the GA content of hazel seeds, in which stringent precautions were taken to minimize the technical problems found by Frankland. All of the pericarp and testa surrounding the embryo was discarded, as these dead tissues had been found to contain substantial amounts of inhibitors (Bradbeer, 1968, and pp. 42–47), although they were not thought to contain GA. Corrections were made for the loss, during the extraction procedure, of GA_3 added at the beginning of the process. A bioassay for GA was chosen in which the GA from the test solution was applied to the pea epicotyl where it specifically promoted elongation (Phillips and Jones, 1964). These precautions increased the sensitivity and reliability of the GA bioassay. In addition, in most cases the GA content of the hazel embryonic axes and cotyledons was determined separately.

As shown in Figure 6.3 and Table 6.3, Ross and Bradbeer (1971*a*) found statistically significant amounts of GA-like activity in hazel embryos. Embryos from freshly harvested nuts contained about 1 nmole of GA_3 equivalent, which fell to about 0.02 nmole on the development of embryo dormancy during dry storage. As found by Frankland and Wareing (1966), there was a very small increase in GA-like material during chilling (50 pmole per embryo as compared to the 0.2 pmole reported by Frankland and Wareing. In the seven days after the temperature was raised to 20 °C, there was a 79-fold increase in the content of GA-like material in the embryo, to 3.96 nmole, a value of the same order of magnitude as that of the amount of exogenous GA_3 known to break hazel seed dormancy (see Figure 6.8).

Thus it seemed that the chilling process brought about changes which permitted the accumulation of GA when the temperature was raised after a sufficient chilling treatment. This implies that the breaking of hazel seed

Table 6.3 Effects of seed treatments on the content of GA-like substances in hazel embryos and on the level of embryo and seed dormancy. After Ross and Bradbeer (1971*a*)

Seed treatments in days			GA content (nmoles GA_3 equivalents)			% of dormant embryo and seeds	
Dry storage after harvest	Chilling at 5 °C	Germination at 20 °C	Embryonic axis	Cotyledons	Whole embryo	Embryo	Intact seed
0	0	0	ND	ND	0.87	23	84
28	0	2	ND	ND	0.02	88	97
28	0	28	ND	ND	0.01	88	97
28	28	0	0.03	0.02	0.05	ND	36
28	28	2	0.80	0.14	0.94	ND	36
28	28	7	2.12	1.84	3.96	ND	36

ND, not determined.

dormancy involves successive exposure to two factors, namely low and higher temperatures, a more complicated situation than already outlined in Table 6.1, in which exposure to a constant single factor, namely chilling, was considered sufficient to break hazel seed dormancy. It may be that the raising of the temperature merely accelerates germination, as prolonged chilling at 5 °C (for more than 12 weeks) also brings about germination at 5 °C, and therefore that only a chilling treatment is obligatory for breaking hazel seed dormancy.

Bradbeer and co-workers followed up this work in a number of ways. The two most abundant substances in the GA bioassays of hazel seed extracts were unequivocally identified as GA_1 and GA_9 by GLC/MS (Williams *et al.*, 1974), although several GA-like substances remain to be identified. In addition to CCC, four other inhibitors of GA biosynthesis were found to inhibit the germination of chilled hazel seeds, thus suggesting that the biosynthesis of GA is a critical post-chilling event in the breaking of hazel seed dormancy (Ross and Bradbeer, 1971*b*). When embryonic axes and cotyledons of chilled hazel seeds were dissected out and incubated separately at 20 °C, GA accumulation in the axes was inhibited by 76% in 10^{-4} M CCC while GA accumulation in the cotyledons was inhibited by only 15% in 10^{-4} M CCC (Arias *et al.*, 1976). On the basis of this and other evidence, the scheme for the breaking of hazel seed dormancy shown in Figure 6.4 was devised. Most of the GA accumulating after chilling is considered to arise from GA synthesized in the embryonic axis, of which about 40% is translocated into the cotyledons during the first 8 days at 20 °C. The growth retardants were considered to have inhibited GA synthesis in the axis. In addition, there appears to have been some GA accumulation which was insensitive to inhibition, and this is thought to represent the release of bound GA mainly in the cotyledons. Approximately 20% of the GA may arise in this way, and it is possible that the

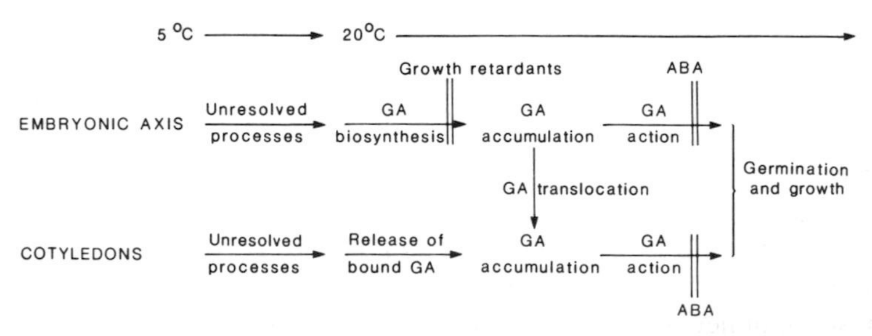

Figure 6.4 Scheme for the breaking of hazel seed dormancy, after Arias *et al.* (1976). Effective growth retardants are CCC, AMO-1618, phosphon D, B-995 and C-011, as described by Ross and Bradbeer (1971*b*).

bound GA was bound during the development of dormancy when the freshly harvested seeds were allowed to dry out at laboratory temperature for 28 days. In Figure 6.4 ABA is shown as an inhibitor or competitor of GA action.

Figure 6.4 provides no more than a fragmentary insight into dormancy breaking in hazel seeds, and it may be that this scheme is quite remote from the critical steps. Virtually nothing is known of the dormancy-breaking reactions which occur during chilling, or of the mode of action of GA, and much remains to be done on the biochemistry of GA biosynthesis. This work represents one of the few reported examples of naturally-occurring and many-fold changes in the concentrations of endogenous plant growth substances, which should justify further research efforts in the future.

In recent years, the development of plant molecular biology has opened up the possibility of pursuing further investigations in this area by the use of DNA probes for the mRNAs coding GA-synthesizing proteins, and antibodies to the proteins themselves.

A number of workers have investigated nucleic acid and protein synthesis in relation to the breaking of seed dormancy by chilling and other means. For example C.A. Anon in King's College London (unpubl.) found that during the chilling of dormant hazel seeds both the embryonic axes and the cotyledons showed an increase in their content of rRNA and poly-A$^+$ RNA (mRNA), as well as a rising rate of incorporation of ^{3}H-adenosine into these fractions. After 6 weeks' chilling at 5 °C, more rapid rises commenced when the temperature of the seeds was raised to 20 °C. Similar increases were observed when dormant seeds were treated with GA_3 at 20 °C so as to break dormancy. When polypeptide synthesis was investigated *in vivo* and by the *in-vitro* translation of extracted poly-A$^+$ RNA, most of the incorporated radioactivity was found to be associated with the major polypeptide bands on SDS-polyacrylamide gels. These major polypeptide bands were presumed to be storage proteins, and this indicated that the embryos (both embryonic axis and cotyledons) had retained the machinery for the synthesis of these substances. As germination progressed the labelling of *in-vivo* and *in-vitro* synthesized polypeptides changed, presumably as a consequence of developmental changes. However, the present state of knowledge is that most newly synthesized polypeptides in germinating seedlings are recognized only as bands on SDS-polyacrylamide gels and that, so far, it has rarely been possible to ascribe a function to such polypeptides. Furthermore, as changes in the qualitative composition of newly synthesized polypeptides have tended to be detected well after the commencement of germination, it seems that up to now no direct evidence has been found to show that the formation of new polypeptides is a necessary part of the breaking of seed dormancy.

It has been necessary to go into some detail on the topic of the chilling of hazel seeds so as to illustrate how little is known about the molecular mechanism of dormancy breaking and how much remains to be done. Space prevents discussion of chilling studies on other species, of which extensive studies on apple seeds in several laboratories in France and Poland should be noted (see, for example, Côme, 1980–81). Similarly, the molecular mechanisms of the other methods of dormancy-breaking are equally obscure, but they will be discussed rather more briefly because of limitations of space.

Dry storage and/or exposure of dry seeds to elevated temperatures

It is quite common for seeds to be dormant when they are fully mature on the parent plant, and for this dormancy to decline during the dry storage of these seeds. Crocker and Barton (1953) listed 42 species with this sort of dormancy and pointed out that they usually showed a state of conditional dormancy in which dry storage brought about a widening of the conditions under which germination would occur. For example, Evenari (1965) quoted the earliest known reference, from 1899, in which Atterberg reported that newly harvested barley germinated at 10 °C but not at 15 °C, and that germination at 15 °C and higher temperatures became possible only after an appropriate period of dry storage. This type of dormancy is a valuable genetic character of cereals in that it prevents sprouting in the ear under the conditions of a wet harvest. Crop breeders would therefore aim for an appropriate level of dormancy which would be broken in time for sowing or malting.

Roberts (1965), in the course of an elegant investigation of dormancy in rice seeds, subjected dormant seeds to dry storage over a range of temperatures. An example of his data may be seen in Figure 6.5 for variety Nam Dawk Nai, in which 50% of the seeds germinated after 38 days of dry storage at 27 °C, while at higher storage temperatures after-ripening occurred more rapidly, so that at 57 °C 50% germination was achieved after less than 2 days' dry storage. Six rice cultivars gave parallel linear plots when storage temperatures were plotted against the logarithm of the number of days of storage required for the achievement of 50% germination. The highest temperature used by Roberts (1965) was overlapped by a series of steam heatings used by Warcup (1980) to simulate the effects of fire or sun on seeds buried in samples of forest soil. The lowest temperature treatments (55 °C for 5–30 minutes) increased the amount of germination of seeds of some species, mainly members of the Juncaceae and

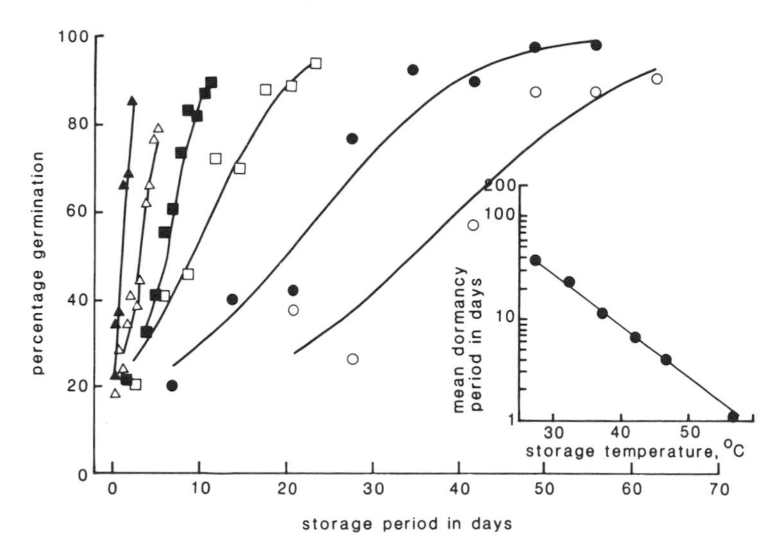

Figure 6.5 The effect of dry storage temperature on the subsequent germination of seeds of *Oryza sativa* L. pureline selection Nam Dawk Nai when the germination tests were conducted in the dark at 32 °C for 7 days. The initial germination was 13%. Storage temperatures: ○, 27 °C; ●, 32 °C; □, 37 °C; ■, 42 °C; △, 47 °C; ▲, 57 °C. The curves were originally fitted by probit analysis. Inset: a plot of the mean dormancy periods over the temperature range 27–57 °C. After Roberts (1965).

Cyperaceae, and brought about the germination of the species which did not germinate in the untreated samples. Further work with forest soils from several sites yielded many species from mainly hard-seeded families, such as the Leguminosae, Geraniaceae and Convolvulaceae, which gave greatly increased germination in soil which had been heated up to 60° or 71°C for 30 minutes. There have been many reports about the role of fire in the maintenance of bush ecosystems in Australia, in which the heat of the fire brings about the breaking of seed dormancy (Beadle, 1940), and also may be necessary for the dehiscence of the seeds from the follicles, as in *Banksia ericifolia* (Bradstock and Myerscough, 1981).

Justification for the inclusion of dry storage and heat treatments in the same dormancy-breaking category is provided by the way that increases in the rate of dormancy breaking result from increases in the treatment temperatures. As the metabolic activity within air-dry seeds is very low, it seems highly unlikely that this mechanism of dormancy breaking involves the removal of a metabolic block. It is more likely that the dry storage or heat treatments damage the coats of hard-coated seeds so that both imbibition and germination become possible. An example is provided by the anatomical investigation of the way in which the strophiole of *Acacia kempeana* lifts and cracks in response to heat, so that entry of water into the seed is facilitated (Hanna, 1984). In addition, it is possible that dry storage and heat may affect the disorganized membranes in the embryos of air-dried seeds.

Light

Evenari (1965) and the authors of the other comprehensive reviews on this topic seem to agree that the promotive effect of light on germination was first reported in 1860 (Caspary), that the first cases of the photo-inhibition of germination were announced in 1903 and 1904 (Heinricher, Remer), and that from 1912 to 1926 Kinzel published lists of species whose germination was promoted by light. The latter worker apparently reported that 672 species out of 964 tested showed induction or stimulation of germination by light. The widespread nature of this phenomenon may be seen from the examples quoted in the next chapter.

Until quite recently it was customary to recognize several groups of seeds as follows:

(i) Positively photoblastic seed in which germination was either induced or promoted by light
(ii) Negatively photoblastic seed in which germination was either wholly or partially inhibited by light
(iii) Apparently non-photoblastic seed in which no difference between dark and light germination had been reported.

Examples of positive photoblastism were abundant, whereas those of negative photoblastism were relatively uncommon. In-depth studies of the mechanism of the light effect have uncovered the phenomenon of photoinhibition of germination, normally in response to prolonged irradiation. This suggests that perhaps for all seeds there may be situations in which light is promotive and situations in which light is inhibitory. The effect of light is dependent on the intensity and duration of irradiation, on the quality of the irradiation (wavelength), on the moisture content of the seed, and on the time of the exposure to irradiation, including the whole of the previous history of the seed during development on the parent plant and subsequently. Imbibed seeds show the greatest sensitivity to light but even air-dry seeds can show a response.

Experimental work with seeds has made an important contribution to our still very imperfect understanding of photomorphogenesis, which is the way that light regulates plant growth and development. Investigation of the promotion of lettuce seed germination by light was critical in the discovery of the photoreversibility of phytochrome, the best known photomorphogenetic pigment of plants (see Borthwick, 1972). The discoveries were made in the United States Department of Agriculture Laboratories at Beltsville, Maryland, largely in the Plant Physiology Laboratory. Borthwick and co-workers used a workshop-built spectograph to determine the action spectra of a number of photomorphogenetic phenomena, including the germination of Grand Rapids lettuce seed, whose germination had been found by Flint and McAlister, in the Seed Laboratory at Beltsville, to be strongly promoted by red light (R) and to be inhibited by far-red light (FR). An action spectrum plotting the amount of irradiation necessary to bring about 50% germination of the light-requiring seeds gave a curve almost identical to those which they had obtained for the induction of flowering in soybean and cocklebur and for the suppression of etiolation in barley and pea. They then supplied seed with sufficient red irradiation to bring about approximately 100% germination, and determined the action spectrum for the light-induced inhibition of germination. The wavelengths giving maximal promotion and inhibition were found to be 660 nm and 730 nm respectively. The physiological investigations led to deductions about the properties of the regulatory pigment which was eventually named 'phytochrome'. Phytochrome was presumed to possess two photoconvertible forms: what seems to be the active form, which predominates after irradiation at 660 nm (the far-red-absorbing form, P_{FR}) and an inactive form which is produced in response to irradiation at 730 nm (the red-absorbing form, P_R).

$$P_R \underset{730\,\text{nm}}{\overset{660\,\text{nm}}{\rightleftharpoons}} P_{FR}\ \text{(active form)}$$

Soon afterwards phytochrome was physically detected by very sensitive spectrophotometry which confirmed its deduced properties (Butler *et al.*, 1959). Purification followed quite quickly, showing it to be a chromoprotein in which the chromophore is a tetrapyrrole. Isolation of phytochrome led to the investigation of its molecular biology, which is still in progress.

Seed germination is one of the many photomorphogenetic responses which are mediated through the agency of phytochrome. The typical light-sensitive seed exhibits what Mohr (1972) describes as the operational criteria of the phytochrome response, in which a short red irradiation stimulates the rate of germination and a subsequent short irradiation with far-red reverses the promotion of germination, as shown by the action spectra in Figure 6.6. In successive irradiations, alternately with red and far-red, it is the final irradiation which determines whether or not

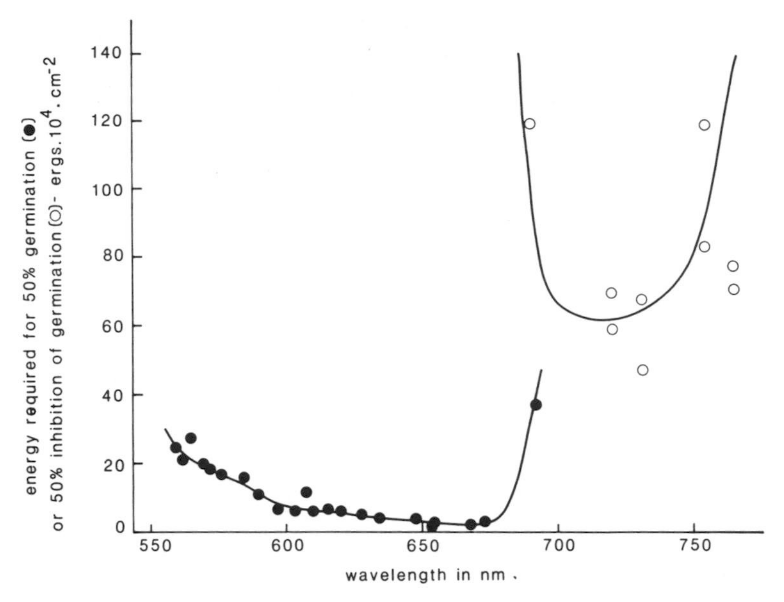

Figure 6.6 Action spectra for the stimulation of the germination of Grand Rapids lettuce seed (●) and for the photoreversal of the stimulation (○). Data of Borthwick *et al.* (1954), redrawn.

germination will be stimulated, as shown in Table 6.4. The photoreversibility of the phytochrome response shows up under conditions of low irradiation or fluence from 10 to 10^3 μE.m^{-2}, as shown in Table 6.5 and illustrated in Figure 6.7. Recent work has shown that some seeds in a population, as for example *Arabidopsis thaliana* (Cone *et al.*, 1985), may be induced to germinate by very low amounts of irradiation, so low that an inductive irradiation can be given very easily by a standard darkroom green safety lamp. This is known as the very low fluence response, and it is apparently not photoreversible.

Although far-red irradiation will normally reverse the promotive effects of low fluences of red light, many photomorphogenetic phenomena may be induced by prolonged irradiation in the high-irradiance reaction in which the far-red and blue are the most effective regions of the spectrum. For the phenomena of seedling photomorphogenesis, both low fluence and high irradiance drive development in the same direction. In contrast, in seed germination, the effects of low fluence and high irradiance seem to be antagonistic in those species which have been investigated; short irradiations of appropriate wavelength are promotive, whereas prolonged irradiations with light of a wide variety of spectral qualities are inhibitory.

Table 6.4 Photoreversibility of the germination of Grand Rapids lettuce seed by alternate irradiations with red and far-red. After 3 h inhibition at 20 °C in total darkness the seeds were subjected to successive irradiations at 26 °C with red (1 minute) and far-red (4 minutes) according to the protocol shown below. The seeds were returned to total darkness at 20 °C and germination was recorded after 2 days. From Borthwick *et al.* (1954).

Irradiations	*% germination*
R	70
R : FR	6
R : FR : R	74
R : FR : R : FR	6
R : FR : R : FR : R	76
R : FR : R : FR : R : FR	7
R : FR : R : FR : R : FR : R	81
R : FR : R : FR : R : FR : R : FR	7

Table 6.5 Levels of sensitivity of the phytochrome response of plants.

Nature of response	Irradiation range of response (μE.m^{-2})	Percentage of phytochrome in P_{FR} form
Very low fluence	10^{-4}–10^{-2}	0.0001–0.01
Low fluence	10–10^3	1–75
High irradiation response	$> 10^3$	1–75

Phytochrome is considered to be the receptor pigment for the high-irradiance reaction, and an explanation of its dual role in the promotion and inhibition of germination has been put forward by Bartley and Frankland (1982) on the basis of experiments with *Sinapis arvensis* seeds, of which an example is shown in Figure 6.7. The low fluence response involves the conversion of sufficient phytochrome into the P_{FR} form. The high-irradiance response depends not only on the amount of P_{FR} but also on the rate of cycling between the P_R and P_{FR} forms. The effects of prolonged irradiation are therefore dependent on both wavelength and fluence rate. Light establishes a photoequilibrium between P_R and P_{FR}, expressed as the relative amount of total phytochrome in the P_{FR} form, the position of the equilibrium being dependent on the light quality. The amount of light is considered to determine the 'rate of cycling'. Despite a very substantial literature on photomorphogenesis, the understanding of the mechanism of action of phytochrome and, consequently, of the way in which light breaks seed dormancy, is very fragmentary. Further consideration of the topic will

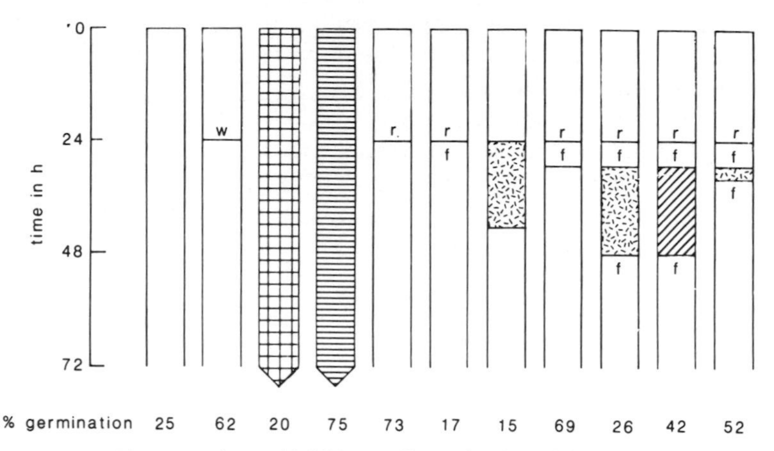

Figure 6.7 The promotive and inhibitory effects of various light treatments on the seed germination of *Sinapis arvensis*. The seeds were irradiated and incubated at 15 °C. Germination was recorded after 72 h. The following treatments and their timing are indicated as follows: *r*, 5 minutes red light; *f*, 15 minutes far-red light; *w*, 15 minutes white incandescent light; unhatched, dark; cross-hatched, high fluence rate white incandescent light; horizontal hatching, low fluence-rate white incandescent light; stippled, high fluence-rate light, $\phi = 0{\cdot}27$; diagonal hatching, low fluence rate light, $\phi = 0{\cdot}27$. After Bartley and Frankland (1982).

be given in the next chapter with respect to the effect of naturally-occurring illumination on dormancy and germination.

Leaching

In a standard petri-dish germination test, compounds can frequently be observed to be leached from seeds. If the leachates contain germination inhibitors, as they do in hazel seeds for example, such leaching must be considered to be a contributor to the dormancy-breaking process. The effectiveness of leaching will depend on the rate of diffusion of the inhibitors from the seed and on the volume and rate of flow of the surrounding water. Very little water may be required, as was found by Wareing and Foda (1957) in their work with the dormancy of the upper seed of *Xanthium pennsylvanicum*. On the basis of conventional germination tests, the isolated embryos had been reported to be non-dormant, but these workers found that if the embryos were imbibed through a saturated atmosphere, rather than by liquid water, no germination occurred. Leachates from embryos were found to contain inhibitors which could cause leached embryos to remain dormant.

The spectacular germination of annuals in hot deserts as a consequence of substantial rainfall is thought to result from the leaching of inhibitors from the seeds. Deserts occur in warm temperate and tropical climates when the average annual precipitation is less than 100 mm. Usually most of the precipitation occurs over a short period of time during the cooler season, and sufficient moisture is retained in the soil for annual plants to complete the germination, growth and reproductive phases of their life cycles. Deserts range from those where sufficient rain falls in virtually every year, to those where the desert may bloom only in occasional years with many dry years in between.

The investigations of F.W. Went and his colleagues on some annual plants of the Californian desert showed that, for seeds sown in moist soil in seed trays, the germination of some species was increased by simulated rain, whereas for other species no germination occurred unless an appropriate treatment with simulated rainfall was provided (Lang, 1965). The amount of germination was dependent on the duration and intensity of the rainfall. The stimulatory effect was clearly that of the leaching-out of inhibitors. At higher rainfall levels germination fell, but an explanation of the nature of the soaking injury which presumably brought this about does not appear to be available.

Scarification

Where seed dormancy is brought about by hard-seededness, in which the seed coverings act as a physical barrier to germination (see pp. 40–42) through the prevention of embryo expansion or radicle growth or through the restriction of water uptake or perhaps of gaseous exchange, the obvious dormancy-breaking mechanism involves damage to the seed coverings. The mechanisms by which scarification is reputed to occur in nature owe rather more to observation and inference than to experimentation. However, one can find suggestions that natural scarification is brought about by such means as trampling by hoofed animals, uncompleted predation by animals (e.g. rodents, birds, insects), damage by fungi and soil micro-organisms, passage through an animal's digestive tract, and extreme changes in temperature. Although all of these mechanisms are feasible and there is little doubt that they occur, some of them probably affect only a minute proportion of the seed population of a species. In hard-seeded species it is more likely that dormancy is broken dynamically by the development of higher hydrostatic pressures within the embryo, as in hazel seeds in response to chilling (see p. 56) and in the seeds of *Xanthium* in response to dry storage (see p. 53). Where the breakdown of the seed coverings occurs in nature, it is probably a result of the slow physical and chemical breakdown of the coverings by oxidation and the absorbance of solar radiation (hard seeds are frequently dark in colour, so that they absorb visible radiation as well as radiation of shorter and longer wavelengths).

Artificial methods have been used with considerable success in breaking the dormancy of hard seeds. In small-scale operations, such as many of those carried out by seed scientists and amateur gardeners, scarification may be carried out manually by such means as the piercing of the seed coat with a mounted needle, or by cutting or nicking the seed coverings with a sharp knife or by abrading the seed with sandpaper. Physical methods of scarification have been mechanized and scaled-up so as to subject seeds to abrasion, percussion or grinding. For example, seedsmen have passed the caryopses of photoblastic lettuce varieties through a roller-mill so that the seed covering is cracked and the seeds can germinate in the absence of illumination. This is intended to reduce the incidence of customer complaint that the lettuce seedlings did not come up (probably because the seeds had been sown rather deeply). A rather more convenient scarification method involves treatment of seeds with concentrated sulphuric acid, normally for anything from 5 to 60 minutes. This operation requires caution and is quite spectacular as the seeds turn black and emit fumes. Excessive exposure will, of course, damage the embryos, and care is necessary to see that the seeds are washed well at the end of the treatment.

Treatment with chemicals

There are a very large number of examples quoted in the literature in which the application of a chemical has been found to bring about the germination of a dormant seed. Some of the chemicals concerned may occur at an appropriate concentration in the natural environment; others may be used in the practice of agriculture, horticulture or forestry. Much of the information concerns the use of chemicals in seed research, since the interaction of seeds and chemicals offers the opportunity of gaining an insight into the mechanisms of embryo dormancy and its removal.

Plant growth substances have been the prime suspects as regulators of dormancy, and, as a generalization, ABA normally imposes seed dormancy (see p. 50) while gibberellins frequently break dormancy, cytokinins are less frequently reported to break dormancy and C_2H_4 may be involved in both the imposition and breaking of dormancy in different species. Figure 6.8 shows the effect of GA_3 on the germination of dormant hazel seeds. It is of interest to see the agreement between the log dose–response curves for the two seasons which were studied. In the intervening season the quality

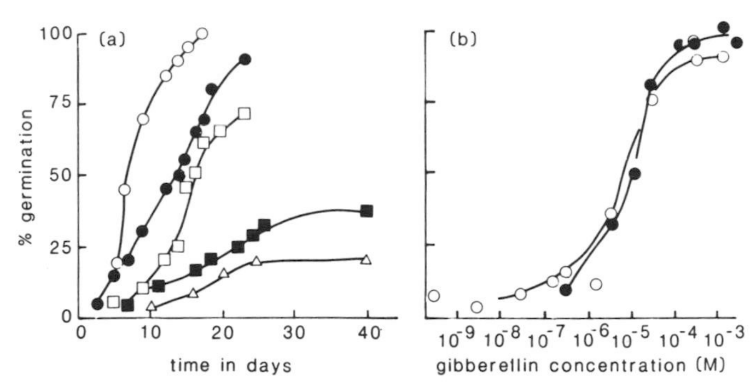

Figure 6.8 The germination of dormant hazel seeds in response to GA_3. (*a*) Germination curves for seeds of the 1961 crop. Gibberellin concentrations: 1.5×10^{-3}M, ○; 3×10^{-4}M, ●; 3×10^{-5}M, □; 3×10^{-6}M, ■; 3×10^{-7}M, △. (*b*) Log dose–response curves for the 1961 (○) and 1963 (●) crops. After Bradbeer and Pinfield (1967).

of the hazel crop was poor and a similar experiment was not completed because of the very high rates of infection of the seeds. Although investigations with plant growth substances have increased our understanding of them and of the physiology and biochemistry of seed dormancy and germination, we remain far from an understanding of dormancy at the molecular level (see pp. 78–79).

As shown in Figure 6.9, E.H. Roberts (1964) found that treatments with a wide range of inhibitors of respiratory electron transport, energy metabolism and intermediary metabolism brought about an acceleration in the rate of germination of rice. These seeds showed conditional dormancy, in that their germination was slow; pretreatment by dry storage and/or elevated temperatures or germination in the presence of the above metabolic inhibitors were necessary for rapid germination. Roberts suggested that the inhibitors affected glycolysis, the tricarboxylic acid cycle and the conventional respiratory electron transport pathway through cytochrome oxidase, and that in some way metabolites were deflected through the oxidative pentose phosphate pathway which somehow accelerated germination. Although these observations may well yield a solution to the enigma of the metabolic basis of dormancy, further investigation of the hypothesis (see for example Roberts and Smith, 1977) have so far yielded equivocal results. Enough has emerged to maintain an interest; for example, when the dormancy of hazel seeds is broken by chilling, there is a relative and absolute increase in the activity of the alternate pathway of respiratory electron transport which is not sensitive to

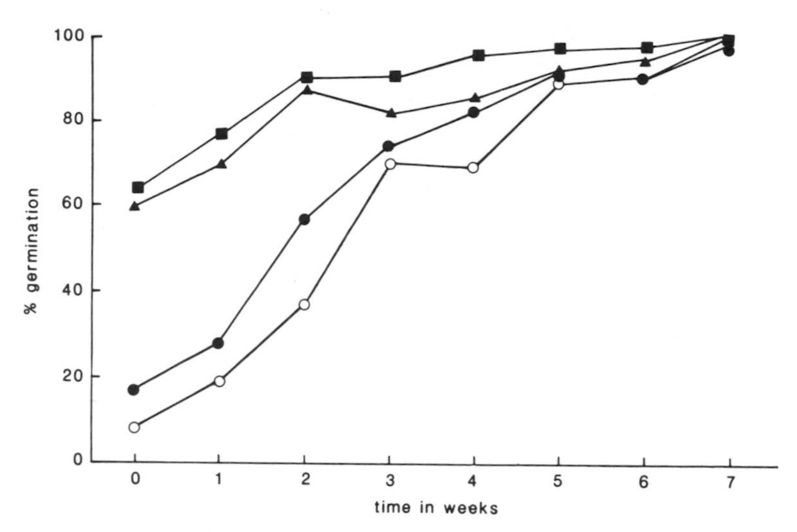

Figure 6.9 The effects of cyanide and azide on the germination of dormant rice seed. A pure-line selection of 'Toma 112' (*Oryza sativa* L. ssp. *indica*) was soaked in the inhibitor, air-dried and subjected to standard germination tests. No pretreatment, ○; pretreatment with water, ●; 10^{-4}M cyanide, ■; 10^{-4}M azide, ▲. After Roberts (1964).

cyanide and the other inhibitors used by Roberts and his co-workers (G.A. Rendon, unpubl.).

When farmers provide a young and growing crop with a top-dressing of nitrate in the spring, a great flush of weed seedlings frequently results. 0.2% KNO_3 is routinely used to break dormancy in viability tests of many of the seeds which are available commercially.

Some chemicals appear to affect the embryo indirectly through effects on the seed coverings, such as scarification by sulphuric acid or hypochlorite or by the extraction of inhibitors or lipids by organic solvents. Organic solvents might have a direct effect on embryos through inhibitor extraction or membrane modification. CO_2 concentrations higher than ambient can build up in the soil and influence both the imposition and breaking of dormancy. O_2 concentrations appreciably above ambient are not expected under natural conditions, but laboratory experiments have shown that high O_2 concentrations may break dormancy in some seeds. Sulphydryl compounds such as dithiothreitol and thiourea, which can reduce sulphur bridges in proteins, are also known to break the dormancy of certain seeds. The many other known dormancy-breaking substances are in some cases natural products, many being secondary plant products, while some are wholly synthetic substances.

Dormancy breaking by fluctuating temperature

Simplicity of experimental design means that the effects of temperature on the germination of dormant seeds would be studied initially by exposure of seed samples to different and constant temperatures. If one of the treatments should break dormancy, the situation in which control and tested samples differ only in one external parameter, namely temperature, might permit the elucidation of the mechanism of dormancy breaking. However, under natural conditions, seeds in the soil are subject to fluctuating temperatures, and it has been observed that, in seeds requiring chilling, random fluctuations in temperature in and above the range of chilling are as effective as continuous chilling at a constant temperature. In addition it has been found that seeds which remain dormant at constant temperature may be induced to germinate by a diurnal fluctuation of temperature. Thompson and Grime (1983) investigated the effect of fluctuating temperatures on the germination of 112 species of herbaceous plants of the Sheffield area of England, which in a previous study had failed to germinate to a high percentage in the light or which responded only to exceptionally high (> 25 °C) constant temperatures. The germination experiments were carried out on a thermogradient bar in which part of the

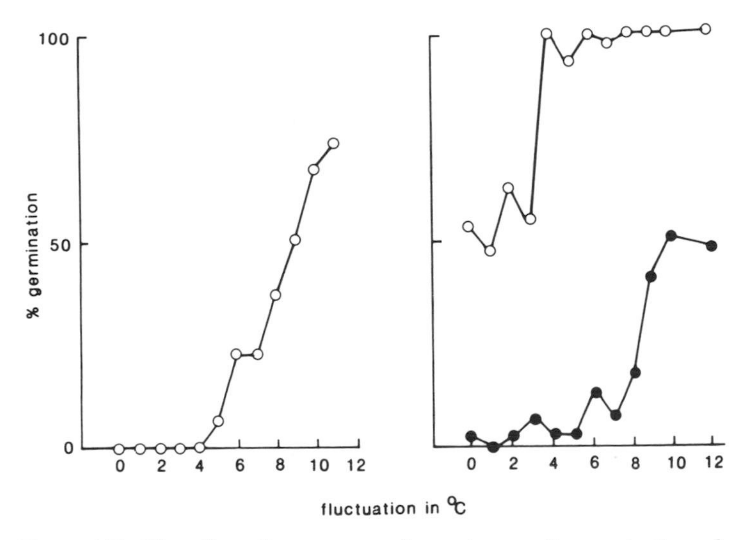

Figure 6.10 The effect of temperature fluctuation on the germination of (*a*) *Rorippa islandica* and (*b*) *Rumex obtusifolius*. The seeds were held at 22 °C for 18 h periods in light or darkness and were subjected to the appropriate temperature reductions in darkness for a period of 6 h every 24 h. ○, 18 hour photoperiod; ●, continuous darkness. After Thompson and Grime (1983).

apparatus was held at a constant 22 °C, with the remainder subjected to a depression of 1–12 °C for 6 h in every 24 h. Forty-six of the species examined were found to have their germination stimulated by temperature fluctuations in the light. Figure 6.10 shows an example of a seed whose germination is dependent upon treatment with light and a fluctuating temperature, and another example in which germination is stimulated by fluctuating temperature in the light and in which some germination occurred in the dark only when the seeds were subjected to a sufficiently large temperature fluctuation.

Temperature fluctuations, as well as fluctuations of day length and light intensity, occur on a seasonal basis. The influence of a seasonally-timed dormancy breaking mechanisms will be considered in Chapter 7.

Dormancy breaking by two or more factors

Many seeds either fail to germinate or show less than 100% germination in response to a single dormancy-breaking treatment of the type outlined in the preceding pages of this chapter. In such cases high germination often results from the use of two distinct dormancy-breaking treatments applied either simultaneously or successively. It is not necessarily feasible, with certain combinations of treatments, for the treatments to be supplied simultaneously. Table 6.6 outlines some results of the application of a range of dormancy-breaking treatments to intact and scarified seeds (actually fruits) of *Rumex crispus* (Hemmat *et al*., 1985). This work shows that light, chilling or a heat shock brought about some germination of the intact seeds and virtually complete germination of scarified seeds. Scarification alone, of seeds held in a germination test in darkness, was virtually ineffective. The application of 0·1 mM mixture of gibberellins A_4 and A_7 had no effect on intact seeds, possibly because the gibberellins were unable to diffuse into the embryo, but gave 43% germination with scarified seeds. Thus combinations of scarification with subsequent illumination or chilling or heat shock or gibberellin treatment were necessary for good germination. The same species had been used for a more intensive study of the effects of stratification, germination, temperature and duration, and illumination, on germination (Totterdell and Roberts, 1979). The lower the chilling temperature, within the range from 15–1·5 °C, the greater the germination, provided that the germination temperature was higher than the chilling temperature. Little *Rumex crispus* germination occurred in darkness, although another species, *R. obtusifolius*, germinated quite well in darkness after chilling. These latter workers found that prolonged chilling brought about the onset of induced or secondary dormancy in

R. crispus. Despite these interesting investigations on *R. crispus*, Baskin and Baskin (1985) claim that the freshly matured seeds of this species are non-dormant and that they do not develop dormancy while they are buried in the soil or attached to the parent plant. However, the germination test used by the latter workers comprised two dormancy-breaking treatments in a 14 h photoperiod at 20 $\mu E.m^{-2}.s^{-1}$ and a set of alternating temperature regimes, so there would appear to be no contradiction in experimental data, as other workers have found that sufficient illumination brings about a high germination of *R. crispus* seeds (Table 6.6). The difference between Baskin and Baskin and the other investigators appears to be in their definitions of dormancy.

The use of combinations of treatments, and their provision over what might be a very long time-scale, gives the investigator the problem of selecting from a very large number of possible experimental protocols, and means that for any species with dormant seed, only a small fraction of the possible combinations of dormancy-breaking treatments have been carried out. Reports of the synergistic effects of two or more factors in dormancy-breaking are abundant; for example, we have found that chilling followed by illumination is quite common, especially in very small-seeded aquatic plants. Hilton (1984) obtained high germination in dormant *Avena fatua* seeds which were supplied with 0.2% KNO_3 and illuminated so as to give a high ratio of FR-absorbing phytochrome to total phytochrome. In dormant hazel seeds combinations of CO_2 and ethylene, light and GA_3 and chilling and light have been promotive. In some cases the effects of two or more factors may be additive or less, in others the results are synergistic, i.e. greater than the sum of the individual treatments.

Table 6.6 Percentage germination of intact and scarified *Rumex crispus* seeds (fruits) in response to different dormancy-breaking treatments. Except where otherwise specified, seeds were dry-stored at 7 °C for 2–8 months prior to treatment. Scarification was carried out in 70% H_2SO_4 for 35 minutes. Except where otherwise stated, the germination tests were carried out at 25 °C for 6 days in darkness. From Hemmat *et al.* (1985).

Treatments	Darkness	Continuous white light 8 $\mu E.m.^{-2}s^{-1}$	5 min red light 7 $\mu E.m.^{-2}s^{-1}$	14 days chilling at 5 °C	1 h heatshock at 40 °C after 24 h[2]	0·1 mM GA_{4+7}[1]
Intact seed	0	74	23	19	18	0
Scarified seed	3	98	77	97	98	43

[1] 7-day germination test
[2] 10-day germination test

The molecular mechanisms of the breaking of seed dormancy

Where dormancy-breaking involves some sort of degradation or leaching from the embryo coverings, it is not difficult to visualize the physical or molecular mechanisms which might bring this about in the laboratory or the field. However, very little direct work seems to have been done on these mechanisms. I am not aware that the precise physical and molecular consequences of the attack of H_2SO_4 on the embryo coverings of any living seed have been investigated in detail, although much early work by natural-products chemists involved the chemical degradation of such products as straw and wood. Nor am I aware of detailed investigations of the mechanisms of microbial degradations of embryo coverings. Seed physiologists are normally content merely to assay whether treatments affect seed germination or not.

The mechanisms involved in embryo dormancy and in the breaking of embryo dormancy in hazel seeds have been considered in Chapter 5 (pp. 48–52) and earlier in Chapter 6 (pp. 58–64). The scheme for the breaking of hazel embryo dormancy shown in Figure 6.4 will be treated as an example of the present state of the elucidation of the mechanism of dormancy breaking. Briefly, chilling is considered to bring about the development of the enzymic machinery for gibberellin synthesis and the release of gibberellins from bound components. On transfer to a higher temperature, gibberellin synthesis and the release of bound gibberellins take place. The gibberellins are then concerned in the mobilization of the seed reserve substances during germination, and in the cotyledonary expansion which is responsible for the rupture of the pericarp. However, the molecular details are almost totally unresolved; only two of the several gibberellins thought to be involved have so far been identified. Virtually nothing is known of the mode of action of gibberellins in hazel seeds or of the synthesis and release of gibberellins subsequent to chilling. The synthetic pathways for gibberellins have not been studied in hazel, and similarly, the way that chilling affects the synthesis of gibberellins is quite unknown. The elucidation of the mode of action of phytochrome or of plant growth substances will probably have to precede a molecular explanation of the mechanism of dormancy breaking. It will be necessary to take account of membrane function and development and of energy metabolism. The study will need to be in depth, as was the case for the currently suspended investigations on hazel seed. With respect to hazel, the possibility of the study of the development of enzymes of gibberellin synthesis and of the relevant DNA and RNA by the use of the techniques of plant molecular biology has begun to open up. The resolution of the mechanism of embryo dormancy breaking will probably result from intensive investigations, even more intensive than those so far pursued with a relatively small number of species such as those listed in Table 5.1.

CHAPTER SEVEN

SEEDS IN THE NATURAL ENVIRONMENT

A perusal of nineteenth-century botanical textbooks and floras reminds us that much of the factual basis of our subject has been quite well established for a long time. For seeds, most of the early information was based on observation and deduction. For example, a great deal was learned about dispersal mechanisms in this way. Experimental investigations developed out of observation, much of it by agricultural research workers concerned with such problems as the enormous populations of weed seeds buried in the soil. In due course some botanists began to see themselves as population biologists, a discipline previously left almost entirely to zoologists. A leading part in the development of this subject area was played by J.L. Harper and his research group, and his classic text *Population Biology of Plants* (1977) includes many of their contributions. That work, and that of J.P. Grime, *Plant Strategies and Vegetation Processes* (1979), include substantial consideration of seeds in this context, specifically with respect to seed production, predation, dispersal, survival in the seed bank of the soil, and germination. These topics are specifically considered by M. Fenner in his recent text, *Seed Ecology* (1985).

Scientific progress nowadays tends to be judged by familiarity with (if not mastery of) expensive apparatus, sophisticated technology and data processing, mathematical models, artificial intelligence and large capital expenditure. The example of scientists like Harper should remind us, however, that the supreme scientific instrument is still the human intellect, and that intellects develop and flourish in the environment provided by teachers, colleagues, students and the whole corpus of human knowledge, written and unwritten. A perusal of any current scientific journal in plant ecology should impress upon the reader the importance of the influence of workers such as Harper and Grime.

Grime adopted the example of animal ecologists by focusing on 'strategies', by which policy a more condensed presentation was permitted. We must dismiss all implications of teleology from this approach; plants do not 'think out' their strategies. Each individual plant inherits the strategy in the form of its genetic composition and the plasticity with which the genome is capable of responding to the environment. It then hands on its strategy, if it attains reproductive success, in the form of its genetic material, variously modified or unmodified by deletions, additions, mutations, resorting and recombinations, to its progeny. Space does not permit an adequate consideration of the principles of plant population biology in this book. The reader is referred to the works quoted in the previous paragraphs.

Figure 7.1 shows an adaptation of one of the two diagrammatic models around which Harper built much of his book. The model outlines the sequence of events by which the success of a seed leads to the production of more seeds. He points out that in this context a plant is merely the means by which a seed produces more seeds, just as a hen is the means by which an egg produces more eggs (Herbert Spencer). To the seed biologist, what could be a more appropriate way of looking at the subject? In Figure 7.1, (I) represents the bank of living seeds lying in a state of true or imposed dormancy in or on the soil. In (II), seedlings are recruited from the seed bank in response to the microenvironmental conditions that occur there. As the demand for the environmental resources by the growing seedlings (III) normally exceeds the capacity of the environment, many of the

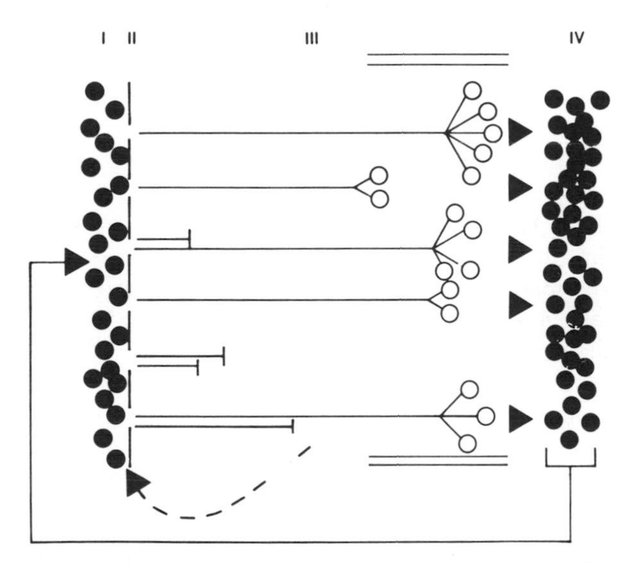

Figure 7.1 Outline of the population of a monocarpic plant. (I) The bank of seeds in the soil; (II) the recruitment of seedlings; (III) the phase of growth in mass and in the number of modular units: (IV) the terminal phase of seed production. Adapted from Harper (1977).

seedlings will die or remain stunted. The surviving seedlings (III) are shown as branching systems in Figure 7.1. These may branch to give a number of modules (tillers, ramets); the potential of branching may be unlimited in theory, although limitations are found in practice as a response to limited resources. During this period of growth, the growing plants change their own environment, and this may effectively alter the recruitment of seedlings from the seed bank; for example, the growing seedlings will cast a shade rich in far-red irradiation, which may inhibit further germination of many species. In the example of the monocarpic plant shown in Figure 7.1, the growth of the plant terminates in the production of a new generation of seeds (IV), which is in turn dispersed into the seed bank to enter the cycle once more. Harper also presents the more complicated model of a polycarpic species in which phases of seed production occur.

The seed bank of the soil

Farmers and gardeners are familiar with the appearance of profuse crops of weeds on bare soil, particularly after cultivation processes, the weeds being considered to result from the disturbance of seeds buried in the soil. In agriculture and horticulture such weeds presented a serious problem in that they reduced the value of the crop and on occasions swamped crops completely. The problem was variously partially counteracted by growing crops in rows so that weeds could be hoed between the rows, by the use of long-strawed cereals and above all by crop rotation, including fallow years and a short-term ley as pasture. The present fashion in the developed world is for the use of selective herbicides to replace good husbandry as a method of weed control. However, weed control continues to be an important area of agricultural research, and it was researchers in agricultural research stations who made some of the first investigations of the weed seed content of the soil—see, for example, Brenchley and Warington (1930); Chippendale and Milton (1934).

The study of soil seed populations requires that soil samples be taken from various depths, either by cutting out quadrats or by the use of a soil borer. Sufficient soil from sufficient samples should be taken to ensure that the error of the determinations be statistically satisfactory. Standard volumes of soil are then taken, sieved to remove stones, and then spread out in seed pans to a depth of 150 mm. The soil should be watered regularly, and every seedling that appears should be identified, recorded and removed. The trays should be kept under ambient temperatures and under cover from rainfall and fresh seed rain. Exposure to sunlight is helpful in breaking the dormancy of seeds which respond to temperature fluctuations, but can be harmful if overheating or drying-out occur. When germination falls off, the soil sample should be given a good turning-over in the pans, and the experiment should be continued. In experiments at the Rothamsted Experimental Station, the soil was thoroughly cultivated every six weeks and turned over every three months. Such procedures may not identify all of the deeply dormant seeds in a soil sample. However, very high populations of seeds have been identified in this sort of way; for example, Brenchley and Warington (1930) estimated that there were 158 million seeds per acre of certain arable land at Rothamsted ($3{\cdot}9 \times 10^8$ ha^{-1}). High seed populations have also been found in habitats which have not been subjected to regular cultivation, such as permanent pasture.

Figure 7.2 is a diagrammatic model of the seed bank of the soil which follows that of Harper (1977). The seed bank is represented by two balloons; the inside one is the dormant seed bank, which contains seeds in a state of true dormancy, requiring exposure to one or more of the dormancy-breaking treatments detailed in Chapter 6, before germination can occur. The outer balloon or active seed bank contains seeds whose dormancy is imposed by the lack of sufficient water for germination or by an unfavourable temperature. After responding to the dormancy-breaking stimulus, those seeds that were truly dormant either germinate immediately, or pass into the active seed bank to await favourable conditions for

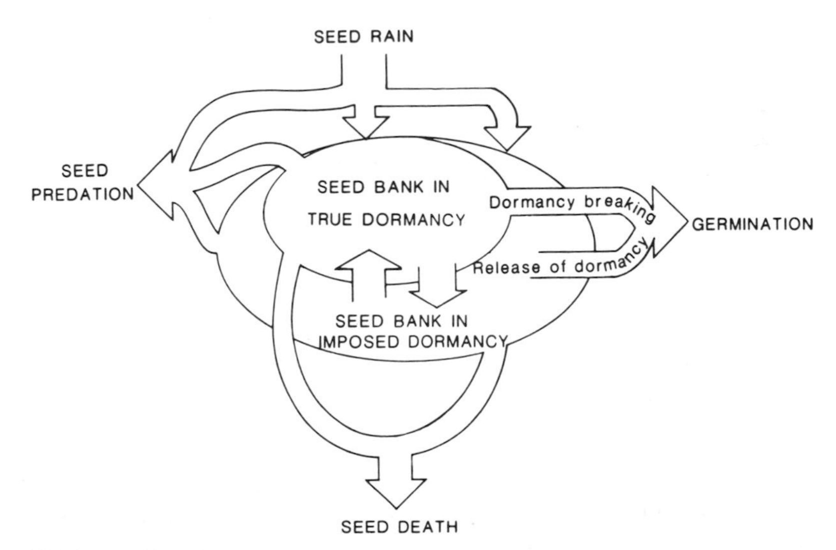

Figure 7.2 Flow chart to illustrate the dynamics of the seed bank, developed from the ideas of Harper (1977).

germination. As a result of their experiences while in the active seed bank, seeds may enter or re-enter a state of true dormancy and become members of the dormant seed bank. Harper presses the banking analogy by likening the active seed bank to a current account and the dormant seed bank to a deposit account. Seed enters each account at the seed bank in the form of seed rain; if it does not leave the bank through germination, the seed will leave through predation or death.

Models, such as those of Harper, have acted as a great stimulus to plant population biologists who have devised many novel observational and experimental approaches to the investigation of seed banks. Direct investigations of the seed banks of most plant species can be made by the use of the binocular dissecting microscope to recognize and remove seeds from soil samples. The capture/recapture techniques used for the estimation of animal populations can be used to provide corrections to the count results. For use with seed banks, a supply of seeds may be marked with a fluorescent dye invisible in visible light. A known number is thoroughly mixed with the soil sample and then the seeds are dissected out of the soil sample for counting. The seeds obtained from the soil sample are then illuminated with a UV lamp so that the recovered seeds fluoresce, and from the recovery rate an estimation of the total number of seeds of that species in the seed bank can be made. An understanding of the seed bank is important in the study of populations of individual species and of plant communities. Such an understanding requires quantitative information about the various fluxes as shown in Figure 7.2. It is necessary to determine the rate of influx of the seed rain, the rate of predation, the longevity of the seeds and their death rates, the factors breaking or inducing dormancy or bringing about germination, and the rates of these processes.

Habitats which are subject to frequent disturbance often possess very large seed banks. In older communities, the seed bank may contain more species than are represented in the community, thus showing more diversity in the soil than at above-ground level. In other words, the seed bank may represent a greater store of genetic material than the growing plants themselves. Although all the seeds of some species in the seed bank may survive for less than a year, the longevity of many species is such that their population in the seed bank represents many growing seasons, which in turn represents something of a genetic memory for the species. Information about the seed bank permits forecasts to be made about the consequences of disturbance of a plant community, such as the possible ability of woodland to regenerate after fire or clear felling. If it is hoped to see natural regeneration of trees which are not represented in the seed bank, then groups of trees should be left to produce seed for regeneration.

The effect of light in dormancy breaking of seeds in the seed bank

Wesson and Wareing (1969) carried out a decisive investigation of the role of light in bringing about the flush of weed seed germination in response to soil cultivation. They chose to look at reasonably long-standing seed banks, under pasture which had been undisturbed for several years but where the land had previously been under arable cultivation. They expected to find that most of the seeds were dormant as a result of their burial. Soil samples, taken from the study fields after discarding the top 2–3 cm to eliminate fresh seeds, were found to give similar numbers of seedlings in germination tests in the greenhouse as did the cultivation of the fields. Similar soil samples were collected from the fields at night under conditions of either total darkness or under the illumination of a 2 W torch bulb behind green filters. Part of the soil was held for germination tests in total darkness and part under light, and it was found that the germination in the dark expressed as a percentage of total germination in light and dark averaged 4.2% for grasses and 7.4% for dicotyledons.

This result was followed up by an interesting experiment in which 75 cm^2 holes were dug by night in the study field to depths of 5, 15 and 30 cm. Three alternate treatments were used: some holes were left uncovered, some were covered by a sheet of glass with soil piled around the edges of the glass, and the remainder were covered with sheets of asbestos, the edges being sealed down with black polythene and soil so as to prevent the entry of light and air. The most striking result (Figure 7.3) is that there was no germination in any of the pits in which total darkness was maintained by asbestos sheets, whereas, for example, about 550 seedlings per m^2 appeared in the course of the 5 weeks that the 5-cm deep pits were covered by glass. Germination was progressively lower in the deeper pits. The pits which were not protected by sheets of glass showed a 30% or more reduction of germination over those that were, presumably in response to the drying of the upper layers of the soil prior to the commencement or the successful completion of germination. When the asbestos covers were removed from the pits after 5 weeks and replaced with glass, much less germination resulted in the following 3 weeks than would have been the case had the newly-dug pits been covered by glass initially. It may be presumed that this fall in germination may be attributed to the effects of the treatment on the deepening of dormancy and on seed predation and death.

The most important of the conclusions reached by Wesson and Wareing was that: 'under natural conditions in the field, the germination of buried seeds, following a disturbance of the soil, is completely dependent upon exposure to light'. The greatest response was to continuous illumination during daylight hours, but the investigators found that exposure to a short

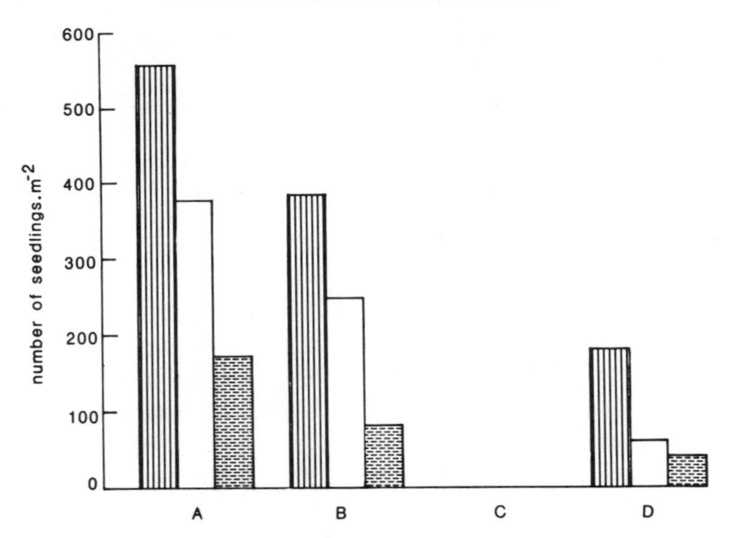

Figure 7.3 The numbers of seedlings found to emerge per square metre from field plots cut to three different depths below the surface. Vertical hatching, 5 cm; unhatched, 15 cm; stippled, 30 cm. (*A*) Plots covered with glass and recorded after 5 weeks; (*B*) plots uncovered and recorded after 5 weeks; (*C*) plots kept totally dark under asbestos covers and recorded after 5 weeks; (*D*) asbestos covers removed from treatment *C* after 5 weeks and emergence recorded after a further 3 weeks. After Wesson and Wareing (1969).

'light break' substantially increased the total number of seedlings emerging from soil otherwise held in darkness. Above all, this work emphasizes the importance of careful experimentation in the field and urges a cautious approach to the extrapolation of laboratory and growth-chamber data to field conditions. It may be that we still underestimate the importance of light in the germination of seeds. Some seeds which were thought not to require light for germination are believed to contain sufficient P_{FR} for germination as a result of irradiation received during seed development and ripening. One of the factors involved in the induction of dormancy during burial in the soil may be the decay of the endogenous P_{FR}. Furthermore, seed biologists may often overlook the effects of very low amounts or levels of irradiation during their work. See the discussion on pp. 66–70.

The effects of seasonal conditions on the germination of seeds in the seed bank

Brenchley and Warington (1930) recorded and presented the numbers of seed-bank seeds which germinated, in their soil samples, on a seasonal basis as they published their data for each quarter of the year in which the records were kept. Thus their records represent germination occurring in winter, spring, summer and autumn. Table 7.1 shows some of their results for soil samples taken from the Broad-balk field at Rothamsted immediately before autumn ploughing in 1925. This field had been continuously cropped with winter wheat since 1843, although problems with weed infestation had necessitated the inclusion of certain patterns of fallow into the rotation between 1889 and 1925. Some seeds germinated rapidly over the first autumn and winter and made no appreciable contribution to the seed bank thereafter. From Table 7.1 it can be seen that the majority of species germinated either in autumn or winter or spring, or divided their germination between autumn and winter or winter and spring. It is interesting to note that the main germination period for *Aethusa cynapium* and *Euphorbia exigua* was in the second winter of the experiment. This may reflect the need for a seasonal cycle before the dormancy of these species is broken. However, it may have resulted from the occasional provision of heat to the experimental glasshouses during the second winter.

Such a broad approach to the seeds in the seed bank has been followed up by a large number of detailed studies of the ecophysiology of the germination of single species or of groups of related species. For example, J.M. and C.C. Baskin have investigated the seasonal basis of germination for a number of species in Kentucky in which they have commenced their investigations with a substantial crop of mature ripe seeds of the study species. A major experiment involved burying fine-mesh nylon bags, each containing about 3000 seeds mixed with soil, at a depth of 7 cm in soil in 15-cm diameter clay pots with drainage holes, the pots being placed in an unheated glasshouse. The pots were watered to field capacity once a week from May to August and daily for the remaining months of the year, except when the soil was frozen. Figure 7.4 shows that buried seeds of *Lamium purpureum* became non-dormant, in that they were able to germinate in petri dishes subjected to 14 h photoperiod and 12:12 h diurnal temperature cycle of 20/10 °C, from June onwards and began to re-enter dormancy in November. Differences in the timing of the entry or departure of the seeds from a dormant state were recorded when different temperatures were used in the diurnal temperature cycle (15/6 °C, 25/25 °C, 30/15 °C and 35/20 °C). Germination was recorded in only three of the 130 germination tests carried out in total darkness, the highest percentage being 4. Thus buried seeds of *Lamium purpureum* are able to germinate only if they are illuminated during late summer and autumn. Similar types of investigation have been carried out with other species in other laboratories.

Table 7.1 The seasonal occurrence of germination in soil from 140 samples taken from the Broad-balk field at Rothamsted prior to autumn ploughing in 1925. The field had been continuously cropped with winter wheat since 1843, although some fallow areas had been used in an attempt to control weed infestations. Each value represents the seed bank in a total soil sample of 35 square feet. The autumn 1925 and winter 1926 values have been added together because of initial difficulties in identifying seedlings during the first three months of the investigation. A, autumn; W, winter; Sp, spring; S, summer. After Brenchley and Warington (1930).

Species	Number of seedlings[1] per 35 ft^2							Germination type
	1925 A 1926 W	1926			1927			
		Sp	S	A	W	Sp	S	
Alchemilla arvensis	5300	350	250	1900	250	40	60	Autumn
Polygonum aviculare	450	220	0	0	1050	30	10	Winter
Myosotis arvensis	600	120	20	80	40	20	0	Autumn/winter
Capsella bursa-pastoris	40	200	50	30	20	50	10	Spring
Atriplex patula	210	410	5	0	110	95	0	Winter/spring
Euphorbia exigua	60	20	10	10	420	20	0	Second winter
Aethusa cynapium	25	15	30	20	275	25	0	Second winter

[1] Numbers taken from graphs in the original publication.

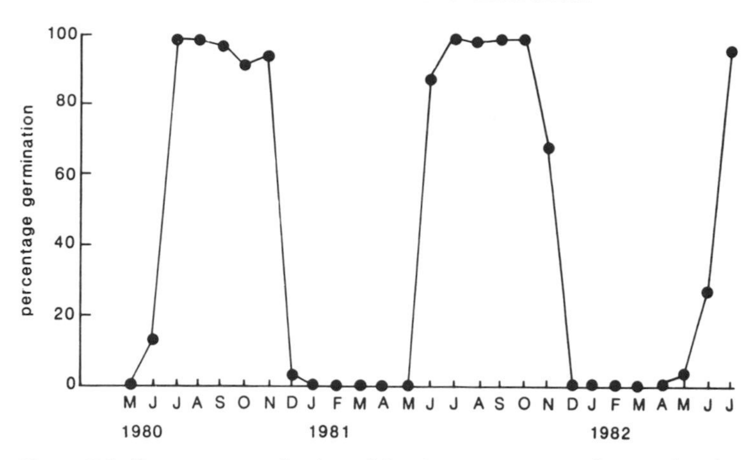

Figure 7.4 Percentage germination of *Lamium purpureum* seeds treated under 14 h photoperiod with a 12:12 h temperature cycle of 20:10 °C following 0–26 months of burial in pots of soil at a depth of 7 cm. Seeds were buried on 13 May 1980 and exhumed on the first day of each subsequent month for a set of germination tests. After Baskin and Baskin (1984)

In a noteworthy study, Sarukhán (1974) investigated the dynamics of the seed banks of three species of buttercup, *Ranunculus acris*, *R. bulbosus* and *R. repens*, growing in the same habitat. He measured the addition to the seed bank in the form of seed rain, the seeds observed to germinate *in situ*, the changing numbers of seeds in enforced dormancy and induced dormancy, seed decay and seed predation. In the case of these three species there was an interesting contrast in that *Ranunculus repens* showed much more vegetative reproduction and much less seed production than the other two species (in 1969 *R. repens* yielded 149.5 seeds per m^2 per year as against 982.5 and 800 respectively for *R. bulbosus* and *R. acris*). The *R. repens* seeds entered a large seed bank in which the half-life was too long to measure with confidence in a two-season study, while the half-lives of the other two species in the seed bank were quite short, 8 months and 5 months for *R. bulbosus* and *R. acris* respectively.

The discussion of seed banks cannot be left without consideration of a very interesting study of the seasonal variation in seed banks of herbaceous species in ten contrasting habitats in Northern England (Thompson and Grime, 1979). These workers set out to identify the species and count the numbers of germinable seeds in the top 3 cm of the soil of each habitat by placing 2-cm deep layers of sieved soil samples over coarse sand in plastic trays. Soil samples were collected at intervals of approximately 6 weeks over the period from October 1974 to October 1975. The germination tests

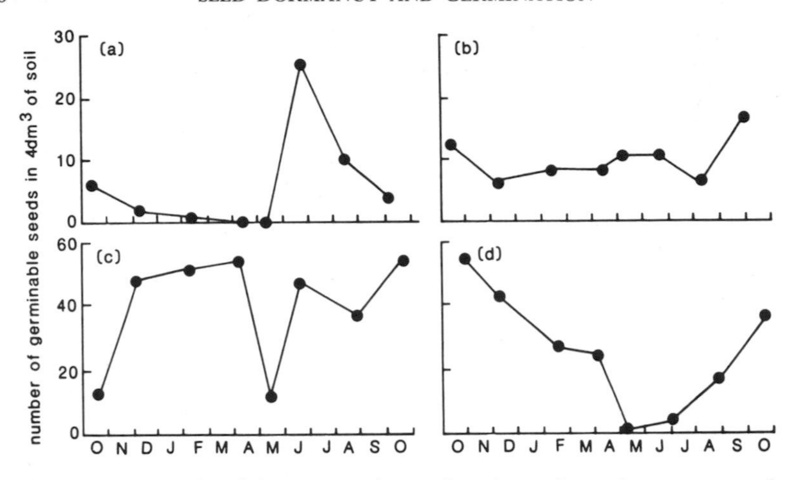

Figure 7.5 Examples of the presence, in samples taken at intervals over a year, of germinable seeds of certain species in the top 3 cm of the soil of certain habitats in Northern England. (*A*) *Festuca ovina* in a derelict herbaceous vegetation on Carboniferous limestone; (*B*) *Thymus drucei* on the same site; (*C*) *Juncus inflexus* on a relatively eutrophic mire; (*D*) *Chamaenerion angustifolium* in a hedgerow. After Thompson and Grime (1979).

were carried out under warm-white fluorescent tubes in a 16 h day at 20 °C and an 8 h night at 15 °C, and the trays were regularly watered from below. Each germination test was carried out for 36 days and care was taken to prevent the induction of further germination by disturbance of the soil. This procedure was designed to detect the seeds in the non-dormant seed bank plus those whose dormancy could be broken by the light or the fluctuating temperatures of the germination test. It was not effective in recovering germinable seeds of such species as *Endymion non-scriptus* in which the fulfilment of the chilling requirement for the breaking of dormancy is followed immediately by the onset of germination (both chilling and germination occur at 5 °C in this species). Figure 7.5 shows as examples one species in which the presence of germinable seed was restricted to late spring and summer, two in which germinable seed numbers remained relatively constant throughout the year, and one in which germinable seeds were present in autumn and winter. It was noted that there was no close correspondence between the relative proportions of species in the germinable seed bank and the species composition of the established vegetation on each site. In several cases, species which were prominent in the vegetation were not detected as seeds, and conversely seeds of some species were recorded for sites where they were not represented in the established vegetation.

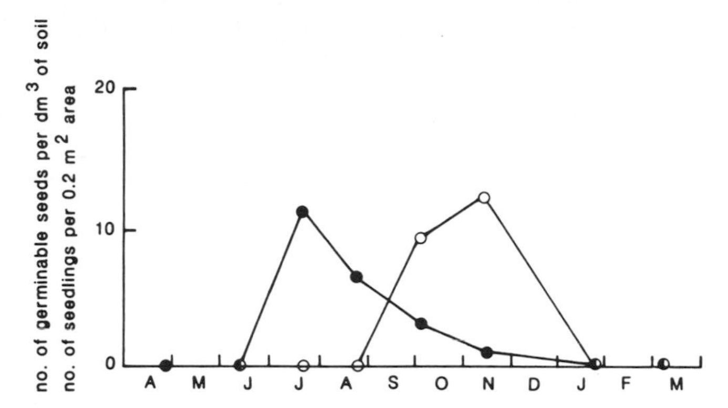

Figure 7.6 Seasonal variation in the number of detached germinable seeds of *Koeleria cristata* in the top 3 cm of the soil (●) together with the number of seedlings germinating on an equivalent area of the same community (○). After Thompson and Grime (1979).

Thompson and Grime (1979) followed up the first part of their investigation with a similar assessment of the number of germinable seeds in the soil of one of the communities for a year, in which the appearance of seedlings in fixed quadrats in the same community was monitored. For example, Figure 7.6 shows that the appearance of germinable seeds of *Koeleria cristata* in late summer and autumn is followed by the appearance of seedlings in the quadrats in October and November.

Among the conclusions reached by these workers was one that the seasonal variation of germinable seed numbers in the soil is a property of the species rather than that of the environment. They recognized four types of seed bank: Type I, of species with transient seed banks present during the summer (e.g. Figures 7.5*a*, 7.6); Type II, of species with transient seed banks present during winter (e.g. *Pimpinella saxifraga*); Type III of species with a persistent seed bank but where many of the seeds germinate soon after they are released (e.g. *Arabidopsis thaliana*); and Type IV of species with a persistent seed bank in which few of the seeds germinate in the period immediately following dispersal and the species maintains a seed bank the size of which changes little with season and is large in relation to the annual production of seeds (e.g. *Calluna vulgaris*). It is not clear to what extent this model will require modification as a result of further investigations and particularly the better understanding of the seed bank dynamics of seeds with various types of deep dormancy. Moreover, as the model applies to a herbaceous species in a number of cold temperature habitats, one would expect to find different situations in both hotter and colder climates, in drier and wetter climates, and in substantially different plant communities.

Laboratory investigation of germination characteristics of the species in the British flora

An account has been published of the germination characteristics of 403 species of local plants studied by the Natural Environment Research Council Unit of Comparative Plant Ecology at the University of Sheffield (Grime *et al.*, 1981). The Sheffield workers performed an initial germination test in a growth room set for 15 h days at 20 °C and 9 h nights at 15 °C. The course of the subsequent investigation depended on the result of the initial test, for example if there was a low germination percentage or a low rate of germination or both, the seed sample was subjected to dry storage, chilling and scarification treatments. If the initial test gave a high percentage germination, the effect of temperature on germination was tested by the use of a thermal gradient bar. The effect of light on germination was tested by germination tests under 'light', 'shade' and 'dark'. If it appeared necessary, the dark treatment was repeated under conditions in which daily germination values were not recorded because of the illumination of the seeds which would occur during the determination of percentage germination.

Since the dormancy and germination characteristics of any species probably represents a very open-ended research problem, such a broad approach can do little more than define questions of interest worthy of deeper investigation for each species. Such an approach also offers the opportunity for making generalizations with respect to different types and groups of plants. For example, on the basis of initial germinability, the Sheffield group arranged the major families into the following series: Gramineae > Compositae > Leguminosae = Cyperaceae > Umbelliferae. The majority of species tested, most markedly the small-seeded species, showed an increase in percentage germination during the course of dry storage. Seventy-five species responded to chilling but the number tested appears to have been only 114 out of the 403 species tested in the programme. Some of the chilled seeds germinated at low temperature in darkness whilst others were dependent on subsequent exposure to light or to higher temperatures or both. Response to chilling was found to be characteristic of the Umbelliferae tested, and the members of the Leguminosae were all found to germinate to a high percentage in response to scarification. These and other generalizations offer challenges which may be tested or pursued by further studies.

Seed production and dispersal

Although this book concerns germination and dormancy, the production and dispersal of seeds should not be totally ignored. Seed size and form have received some attention in Chapter 3. Plant population biologists (see p. 80) have given much consideration to the strategies of plant reproduction, including the proportion of resources devoted to seed production and the strategies of seed dispersal. They attempt to provide rational explanations for plant survival during the course of evolution. Of course all existing species have been evolutionarily successful so far, although one might regard an abundant species dominating a common plant community as being *highly* successful, as for example some of the weeds of cultivation, the abundant species of upland communities and the dominant trees of climax forests. Nevertheless, a plant which is uncommon because it occurs in a rare environmental niche may also be described as successful within the limits of adaptation to an unusual niche. However successfully a species is adapted for survival, the overwhelming influence of man in the exploitation of global resources provides a potent force for species extinction.

The seed provides the means by which a species may colonize sites suitable for its growth. Harper (1977) defines such sites as 'safe sites' in that they provide all of the requirements for germination and seedling establishment as well as a freedom from predators and disease-causing organisms. The seed morphology of the majority of species shows features which may be regarded as dispersal mechanisms. The mechanisms may rely on the parent plant to catapult the seed clear of the parent, or alternatively seed may be dispersed with the aid of some external agency. Seeds may be adapted to wind dispersal by the modification of the testa or pericarp to form some sort of wing to lengthen the dispersal pathway of the seed, or by the lightness of some seeds (5–10 μg) which allows them to be blown for very considerable distances. Some seeds float and may be dispersed by water, a mechanism which permits reed-swamp plants to colonize newly exposed areas of mud. Animals assist in the dispersal of many seeds by carrying them either internally or externally. In internal transport, the animal ingests seeds, and those which survive digestion may subsequently germinate in the pellet or the faeces. In external dispersal, seeds which lack apparent adaptation for dispersal may be carried in mud on the feet of animals (including birds). Alternatively, seeds may be adapted for attachment to animal coats and dispersed in that way. In some plant communities ants are responsible for seed dispersal. Usually the ants are attracted by the oil contained in the elaiosome. They may carry seeds as much as 70 m, remove the elaiosome and leave most of the seeds within the nest under conditions favourable for germination and unfavourable for predation. In Britain this phenomenon is common amongst mesophytic species growing in woodlands, such as *Lamiastrum galeobdolon* and *Viola odorata*. In Australia it was recognized in about 1500 xerophytic species of dry heath and sclerophyll vegetation (Berg, 1975).

The present very brief consideration of seed dispersal has so far concentrated on dispersal over relatively short distances of from several centimetres to several hundred metres. At such a scale, an experimental approach can supplement the observational one. An even more intriguing question concerns single-step dispersal over many kilometres or hundreds of kilometres. It has been deduced that most of the vegetation of newly formed oceanic islands probably arrived in the form of seeds carried within the alimentary canals of migrating birds (Carlquist, 1967). For further information about seed dispersal, see Murray (1987) or the ecologically oriented works mentioned at the beginning of this chapter.

CHAPTER EIGHT

SEED VIABILITY AND VIGOUR

Viability

A viable seed is one which is capable of germination under suitable conditions. The definition includes dormant but viable seeds, in which case the dormancy must be broken before viability can be measured by germination. A non-viable seed, therefore, is one which fails to germinate even under optimal conditions, including treatments for the removal of dormancy. The practical definition of viability depends upon the context in which it is used; for example, to the ecologist, viability implies the ability of the seed to germinate and the ability of the seedling to establish itself in the environment in which the seed finds itself. However, when seeds are utilized, to produce a crop, for example, then viability is a measure of the suitability of the seed batch to produce a satisfactory crop.

In order to maintain confidence in their products, commercial seedsmen apply rigorous criteria of quality control. Some of these criteria, including viability, have legally enforced minimum standards for agricultural and horticultural crops (see Chapter 9). In the UK these legal requirements are effective, and the present writer has never been supplied with commercial seed which fell below these standards. Seeds of many ornamental species are regarded as 'difficult', often because they are dormant but sometimes because they readily lose viability.

Legal and commercial requirements demand that there must be rigorous consistency in the determination of seed viability. Under the International Rules for Seed Testing (International Seed Testing Association, 1985), precise criteria are defined for testing the viability of all of the main species of crop plants and of many trees. They include the criteria necessary for making a representative sampling of a seed batch, whether it be a bag of seed or a trainload, for the number of test replications, and for the precise conditions of the test for each species. These conditions specify the amount of water necessary, the air temperature, and, through the specification of the substrate, the aeration, the amount of contact between seed and moisture, the seed temperature and the illumination. The germination tests may be conducted in dishes with the seeds placed on a specified grade of germination paper as described on pp. 3–5. Some seeds should be placed between sheets of paper; larger seeds may be placed on sand under sheets of paper, pressed into sand or buried in sand. The dishes of seed may be placed in incubators or plant growth cabinets and are normally subjected to continuous illumination. Incubators specifically modified for seed germination tests are available commercially, as are other items of apparatus such as the Copenhagen tank, which provides moisture to the seeds continuously and which would be held in a controlled-temperature room under natural or artificial light. Table 8.1 gives some examples of the standard germination conditions. In a viability test, the duration of the test is also specified, at the end of which the percentage viable seedlings is recorded. A viable seedling must have a strong root system, a shoot and a sufficiently developed leaf system to ensure autonomy. It is usual to record numbers of viable and non-viable seedlings, and if there are appreciable numbers in the latter category, to note the nature of the lesions involved.

Laboratory viability tests of this sort almost invariably give higher germination values than tests conducted in soil, either in seed trays in a glasshouse or under natural conditions in the field. The main cause of the discrepancy is the variability of the available moisture in the soil, both spatially within the soil and temporally as a result of evaporation and drainage on one hand and precipitation and irrigation on the other. Other possible causes of reduced germination in soil are predation, the presence of soil-borne pathogens, insufficient light for optimal germination, and unsatisfactory temperatures. A viable seed should be able to germinate and give rise to a healthy seedling in the presence of its own microflora and that of the medium. However, it is not desirable to expose the seeds to unnecessary micro-organisms and, in addition, infected seeds or seedlings should be removed so as to prevent the spread of infection to healthy material. Even when soil germination tests are carried out in pots in an incubator, the apparent viability may be less than that found in the standard germination tests, as may be seen in Figure 8.1*a* showing data obtained in an investigation of the maintenance of jute seed viability. The figure shows that there was a greater loss of viability at higher seed moisture contents, and that at the higher levels of viability there was only a small difference between the results of the two tests, whereas at the lower levels of viability the soil test gave much lower values.

There are other situations where seed viability is critical, as for example in malting where one of the key qualities of a malting barley is its ability to germinate rapidly and synchronously in the malthouse. Current practice in malting for brewing in the UK is to add 1 g of gibberellic acid per tonne of

Table 8.1 Outline examples of standard germination conditions recommended by the International Seed Testing Association (1985) for the determination of seed viability. The exact procedures are described in that publication.

Species	Substrate[1]	Temperature (°C)	First count (days)	Last count (days)	Additional recommendations including directions for breaking dormancy[2]
Beta vulgaris	TP; BP; S	20–30; 20	4	14	Prewash (multigerm: 2 h, genetic monogerm: 4 h)
Brassica oleracea	TP	20–30; 20	5	10	Prechill; KNO_3 (0·2%)
Corylus avellana[3]	S	20; (20–30)	14	35	Remove pericarp and prechill 2 months at 3–5 °C
Daucus carota	TP; BP	20–30; 20	7	14	—
Festuca rubra	TP	20–30; 15–25	7	21	Prechill; KNO_3 (0·2%)
Hordeum vulgare	BP; S	20	4	7	Preheat (30–35 °C); prechill; GA_3
Lactuca sativa	TP; BP	20	4	7	Prechill
Lycopersicon esculentum	TP; BP	20–30	5	14	KNO_3 (0·2%)
Phaseolus vulgaris	BP; S	20–30; 25; 20	5	9	—
Pinus sylvestris	TP	20–30; 20	7	21; 14	Eastern Mediterranean provenances may require prechill 21 days at 3–5 °C
Pisum sativum	BP; S	20	5	8	—
Triticum aestivum	TP; BP; S	20	4	8	Preheat (30–35 °C); prechill; GA_3
Zea mays	BP; S	20–30; 25; 20	4	7	—

[1] BP (between paper), germinated between two layers of paper; TP (top of paper), germinated on top of one or more layers of paper; S (in sand), seeds planted in a uniform layer of moist sand and then covered to a depth of 1–2 cm with sand which is left loose.
[2] Illumination generally recommended. Precise instructions about illumination or darkness are given if appropriate.
[3] Topographical tetrazolium test preferred.

barley so as to obtain a controlled, synchronous and rapid sprouting of the barley.

Vigour

The measurement of seed viability as described above does not on its own provide sufficient information to estimate the chances of the successful field establishment of the seedlings arising from a particular seed batch. Germination must be sufficiently vigorous for seedling establishment to occur, hence the requirement for the concept of seed vigour. There have been many attempts to define vigour and to agree on a test for vigour (see Heydecker, 1972), and indeed the International Seed Testing Association germination tests have evolved into a measure of both vigour and viability through the assessment of the potential of seedlings to give rise to healthy plants. It is widely considered that the rate of germination, as determined for seeds tested in a standard germination test, provides a satisfactory direct measure of vigour. In jute, for example, good newly harvested seed will normally show germination by radicle emergence within 24 h of the commencement of a germination test on filter paper in a petri dish at 30 °C, and all viable seeds will germinate within four days. Khandakar and Bradbeer (1983) modified the expression used by Jain and Saha (1971) as follows:

$$\text{vigour value} = \frac{\frac{a}{1} + \frac{b}{2} + \frac{c}{3} + \frac{d}{4}}{S} \times 100$$

where *a*, *b*, *c* and *d* respectively represent the number of seeds which germinated after 1, 2, 3 and 4 days of imbibition at 30 °C, and *S* represents the total number of seeds to have germinated. Figure 8.1*b* shows the result of the application of this method to the same batches of jute seed for which viability tests had been conducted. It shows parallel losses of viability and vigour with increasing moisture content. Each test in Figure 8.1*a* and *b* was carried out with 100 seeds.

Figure 8.1*c* shows the results of tests in which 20 seeds from each treatment were germinated in a petri dish under continuous illumination (33 $\mu E.m^{-2}.s^{-1}$) for 4 days at 30 °C, at which time the mean lengths of the shoots and primary roots of the germinated seedlings were recorded. Root growth, but not shoot growth, proved to be a good measure of vigour; indeed, if germinating seedlings fail to develop a strong root system, their chances of survival are substantially reduced. Loss of viability and vigour is also correlated with membrane deterioration, and sugars and other

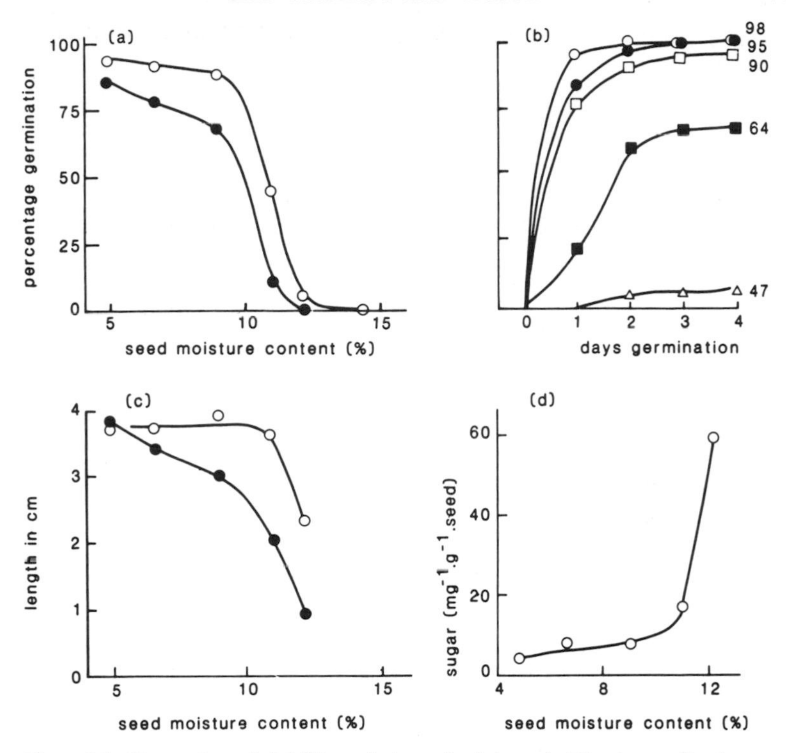

Figure 8.1 Parameters of viability and vigour for jute seeds (*Corchorus olitorius* cv. 0-4) which had been harvested in November 1976 and adjusted to a range of moisture contents in April 1977 before being stored in sealed containers at 32 °C for 2 years. The tests were carried out in April 1979. (*a*) Germination at 30 °C recorded after 4 days in a petri dish (○) and 12 days in a pot of soil (●). (*b*) The course of germination in a petri dish at 30 °C; the figures on the right represent the vigour value (see p. 98); ○, ●, □, ■ and △, initial moisture contents of 4.9, 6.7, 9.0, 11.0 and 12.2%, respectively. (*c*) Shoot (○) and primary root (●) growth after 4 days under illumination at 30 °C (only those seeds which germinated were measured). (*d*) Sugar exudation during 24 hours' elution at 30 °C. After Khandakar and Bradbeer (1983).

compounds are exuded during imbibition. In Figure 8.1*d*, sugar exudation from jute seed was measured by the anthrone technique. This method can identify the more seriously deteriorated jute seed samples, for which the sugar (glucose) can be assayed semi-quantitatively with the reagent-impregnated sticks used for the detection of sugar in urine.

For jute seed, the petri-dish test of viability, together with the assessment of vigour from either the rate of germination or root growth, have been found to give an acceptable estimate of the suitability of a seed batch for crop production, at least with respect to the parameters of viability and vigour. The other parameters of seed quality are considered in Chapter 9. The sugar exudation test can be used to pick out seed batches in which vigour is considerably reduced. Such a generalization holds true for dicotyledons, but seedling growth is more difficult to measure in monocotyledons, where the leaf bases enclose the shoot and where the primary root is overtaken by the secondary roots.

An alternative approach to the assessment of viability and vigour is the use of vital stains such as tetrazolium chloride to stain those cells, and incidentally any contaminating micro-organisms, which possess dehydrogenase activity. The treatment depends on the species, but the seed must be fully imbibed before being subjected to about 24 h soaking in a 1% aqueous solution of 2,3,5-triphenyl-tetrazolium chloride, pH 7, in darkness. The living cells develop a red precipitate, whereas any dead cells remain colourless. The positions of any areas of dead cells will determine whether the seed is capable of germination, and critical examination of the stained seeds can provide an indication of the vigour of the seed batch. In skilled hands, the results of vital staining can be similar to the results of germination tests for viability and vigour. For instance, if more than 90% of a seed batch gave complete and intense staining, then there could be confident expectations of acceptable crop establishment. However, there is no real substitute for germination testing, since the presence of some adverse factors, such as phytotoxic seed dressings or the spores of pathogenic micro-organisms, would not be detected by the tetrazolium test. The tetrazolium test has its uses; for instance it will give a positive result for a dormant seed and, for those parts which fail to stain, may provide evidence of the cause of the lesions suffered by the seed.

At the cellular level, vigour and viability depend on the integrity of the cellular machinery, of the DNA of nucleus, plastids and mitochondria, of DNA replication and transcription, of protein synthesis and of the integrity of organelles and of other cellular membranes. In dry seeds, membranes become somewhat disorganized so that their re-assembly during imbibition is an essential prerequisite for the onset of germination.

Factors affecting the maintenance of viability and vigour during storage

This is a key consideration in seed technology (see Chapter 9) with respect to the relatively short-term storage between harvest and sowing time or the longer-term requirements of a genetic resource seed bank. The three main parameters of the storage environment which determine seed survival are seed moisture content, temperature, and atmospheric oxygen concentration. Roberts (1972) summarizes the results of a great number of investigations by stating: 'In the vast majority of cases it has been shown that the lower the temperature and the lower the moisture content the longer the period of viability'. There are exceptions, as Roberts points out, in that some seeds have a relatively high optimum moisture content for maximum retention of viability, such as the larger-seeded temperate hardwoods (e.g. *Aesculus*, *Corylus*, *Fagus* and *Quercus* species) and many species of warmer climates (e.g. *Caffea* and *Citrus*), such species being described as recalcitrant, because long-term storage of seeds at high moisture contents presents an intractable problem. Apparent loss of viability at low moisture contents can result from the death of seeds during the drying process. For instance, when jute seeds (*Corchorus capsularis* cv. D-154) were dried from 9.8% to 5.8% moisture in an incubator at a relative humidity of 20% over the course of 2 days at 32 °C, there was an immediate fall of viability from 100% to 77%. Subsequently, during 35 months' storage in sealed lamofoil pouches at 32 °C, viability fell to no more than 60%. As a more accurate generalization, it might be better to say that each seed batch will have an optimal moisture content for the maintenance of viability and vigour. A similar statement may well be appropriate with respect to temperature and oxygen concentration, with the rider in both of these cases that for some seeds the optimum may be the lowest temperature or oxygen concentration that can be achieved.

Moisture content

Figure 8.2 shows the results of an empirical determination of the effects of initial moisture content on the viability of jute seed (*Corchorus olitorius*) stored in sealed containers at 32 °C. Seed survival (or seed death) is shown in the form of a sigmoidal curve for each treatment. Each curve represents the variation between the survival periods of the individual seeds in the seed sample. Except for seeds held under conditions in which very rapid loss of viability occurs, the periods of viability of individual seeds occupy a normal distribution around a mean value (the mean being the time required for 50% loss of viability). Such a normal distribution will result in a straight-line relationship when percentage germination is plotted on a probability scale against time, as has been done in Figure 8.3 for the data of Figure 8.2. Mathematical analysis of survival data of this type has permitted the construction of nomographs for the prediction of viability for seeds held under hermetic storage conditions (Roberts, 1972).

In the experiment shown in Figures 8.2 and 8.3, 200 g jute seed samples were sealed in lamofoil pouches which were unsealed and promptly resealed every two months for the removal of samples for germination tests

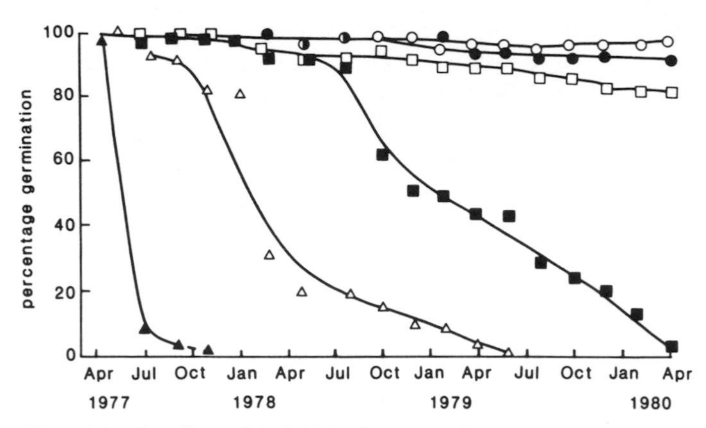

Figure 8.2 The effects of the initial adjustment of seed moisture content on the maintenance of viability in jute seeds (*Corchorus olitorius* cv. 0-4) from the 1976 harvest. The seeds were obtained with a moisture content of 11.0% which was initially adjusted to 4.9% (○), 6.7% (●), 9.0% (□), 11.0% (■), 12.2% (△) and 14.3% (▲) by incubation at 32 °C at 100% or 20% relative humidity as appropriate. Storage was carried out at 32 °C in sealed lamofoil pouches. Each point represents a petri dish germination test carried out with 100 seeds (4 petri dishes) for 4 days in darkness. After Khandakar and Bradbeer (1983).

and moisture content determinations. Storage at 32 °C in these pouches permitted a slow fall in seed moisture content; for example, in three years, the sample with 11.90% moisture content fell to 8.1% and the one with 9.9% moisture fell to 7.2%. In this case initial moisture contents of 4.9% and 6.7% permitted the maintenance of better than 90% viability for three years. The moisture content of a seed batch can be maintained constant by hermetically sealing up the seeds. In open storage, the moisture content of the seeds will change as a result of the equilibration of the moisture contents of seeds and atmosphere.

Temperature

Viability and vigour decline more rapidly at higher temperatures and less rapidly at lower temperatures, as shown in Figure 8.4 for the decline in viability of rice seed stored in a range of temperatures between 27° and 47 °C. The theoretical curves for the fall in viability were constructed by Roberts (1961*b*) and can be seen to provide a good fit for the data, within the limits of experimental error. From Figure 8.4, the number of days required for viability to fall from 100% to 25%, determined and plotted in Figure 8.5 on a logarithmic scale against temperature, shows a log-linear

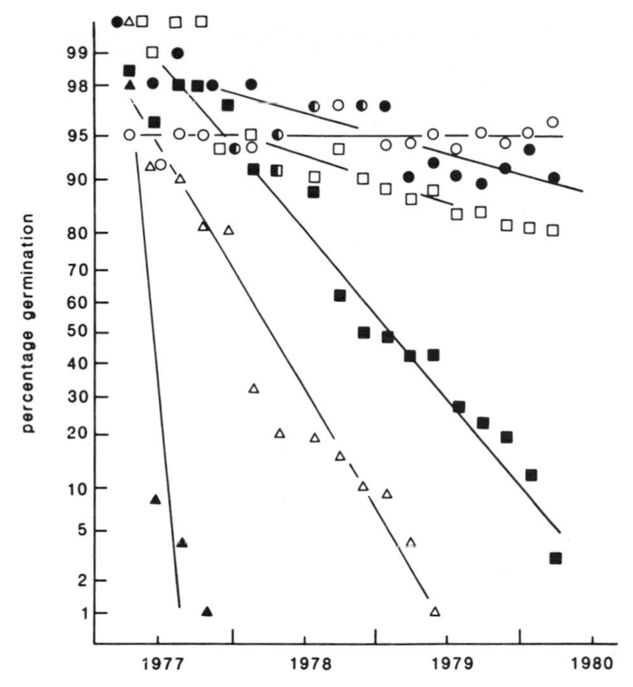

Figure 8.3 The data from Figure 8.2 expressed with a probability scale as ordinate.

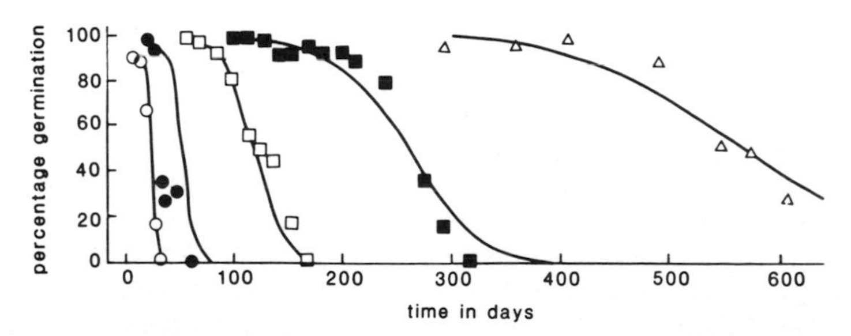

Figure 8.4 The effects of temperature on the maintenance of the viability of rice seed (12% moisture content) hermetically sealed in air. Storage temperatures: △, 27 °C; ■, 32 °C; □, 37 °C; ●, 42 °C; ○, 47 °C. After Roberts (1961*b*).

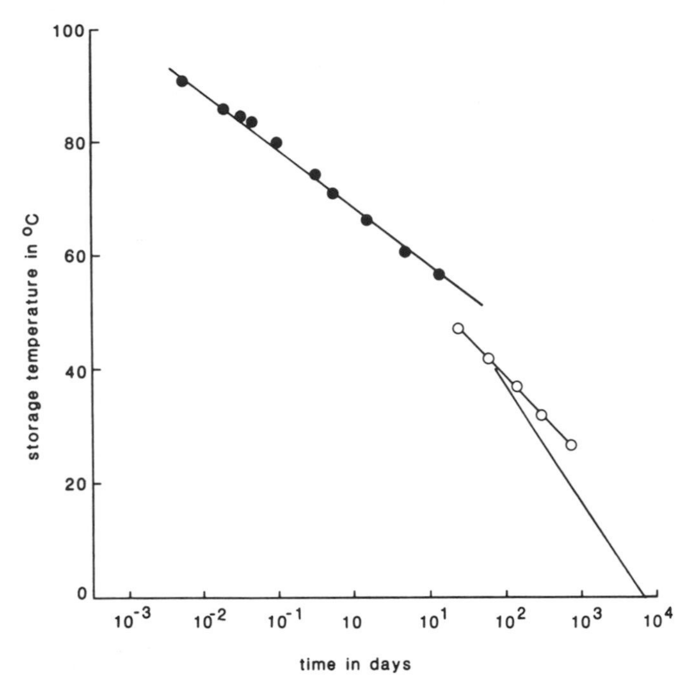

Figure 8.5 The relationship between storage temperature and the time required for seed viability to fall to 25%. Data for Turkish Red wheat with 12% moisture content, ● (Groves, 1917); rice variety Toma 112 with 12% moisture content, ○ (Roberts, 1961*b*) and calculated values for wheat of 12% moisture content.——(Roberts and Roberts, 1972).

relationship. Also included in Figure 8.5 are two similarly plotted sets of data for wheat, one determined by Groves (1917) for temperatures between 56 °C and 9 °C, and a line derived from a viability nomograph constructed by Roberts and Roberts (1972) for temperatures between 0° and 40 °C. All three sets of data give log-linear plots from which the temperature coefficients (Q_{10}) for the loss of viability have been determined to be 9.2 for wheat in the high-temperature range and 3.0 in the low-temperature range and 4.7 for rice. Extrapolation of these lines to a 0 °C storage temperature shows that wheat viability was predicted to fall to 25% after 17.2 years at 0 °C according to the data of Roberts and Roberts (1972), a result which is in accord with the known longevity of wheat seed. In contrast, the data of Groves may be extrapolated to predict a viability period of about 3000 years at 0 °C, a value grossly in excess of actual longevity, which shows that rapid ageing at high temperatures may not provide a reliable guide to the maintenance of viability at normal storage temperatures.

Provided that the moisture content of a seed sample is sufficiently low, seeds frequently can maintain viability when stored at temperatures below 0 °C. Temperatures appreciably below 0 °C are satisfactory for storage, as a rule, although too rapid freezing and thawing can result in physical damage, such as the fracture of some organs of the embryo. Commercial deep-freezes held at about − 18 °C are frequently used for seed storage.

Oxygen concentration

In experiments designed to study the effects of constant conditions on the maintenance of seed viability, it is most convenient to prevent fluctuations of seed moisture content by placing the seeds in a sealed container. Under such conditions, the respiration of the seeds and of their associated microflora results in a fall in the oxygen concentration and a corresponding rise in that of carbon dioxide. The lower partial pressures of oxygen favour the maintenance of viability, and equally good results have been obtained by replacing the air in storage containers with nitrogen, carbon dioxide or argon. For some, but not all, seeds, storage in a vacuum is equally effective. When Abdalla and Roberts (1968) compared the maintenance of viability of seeds of barley, broad bean and pea under nitrogen, air and oxygen at 25 °C and highly unfavourable moisture contents, they found that air and 100% oxygen gave a much greater loss of viability than nitrogen and that the loss under 100% oxygen was little greater than that in air.

The mechanism of ageing

Despite considerable scientific interest the processes of ageing, senescence and death are poorly understood. Of the many hypotheses put forward with respect to seed (see Roberts, 1972), those relating to effects extrinsic to the seed, such as ionizing radiation or the effects of storage fungi, seem closer to resolution at the present time. The other hypotheses concern cases where death may come about as a result of events occurring within the seed, i.e. intrinsic events, in which damage to cells occurs. Although some of the damage may be repaired within the cell, the accumulation of defects may eventually lead to cell death, and, when cell deaths outstrip the resources for the production of new cells, organ or organismal death ensues.

Several types of damage have been shown to be correlated to the loss of seed viability and vigour. Older seeds are known to produce more genetic variation than younger seeds and this has been correlated with an increase in visible chromosomal damage during storage. Furthermore, chromosomal damage is promoted by the same storage conditions that favour loss of viability, namely high moisture, temperature and oxygen. Since germination up to radicle emergence can occur in the absence of cell division, it is unlikely that chromosome damage can be the prime cause of loss of viability, although damage to DNA undoubtedly reduces seedling vigour. DNA repair is an essential survival mechanism of living organisms, but although DNA damage occurs in the unimbibed seed, the repair mechanisms cannot recommence until imbibition commences. For example, Osborne and co-workers have detected DNA repair during the earliest stages of embryo imbibition (Osborne, 1982). Roos (1982) points out that chromosomal aberrations appear to be largely eliminated during the growth of the plant, and thus are not passed on to the next generation.

Osborne (1982) defines loss of vigour in biochemical terms as a progressive loss in potential to synthesize proteins, lipids and RNA. The synthesis of these macromolecules commences within minutes of the imbibition of dry embryo cells, but the first round of DNA replication always occurs late in the reactivation sequence. At the lethal limit, the stage of DNA replication is never reached. Osborne defines the final stage in loss of viability in dry-stored seeds as the complete absence of all synthetic activity on imbibition, the embryo being pronounced 'dead'. However, long before this stage the seed samples would have reached zero viability, the stage at which no radicle growth could be detected.

Other symptoms of ageing in seeds are loss of enzyme activity and the accumulation of membrane damage. Simon (1984) has pointed out that in dry seeds there will be insufficient water for the membrane bilayers to maintain their integrity and that consequently during the early stages of imbibition the reassembly and repair of membranes takes place. In ageing seeds, recovery is incomplete, and damage to all types of cell membrane can be found (see, for example, Berjak and Villiers, 1972).

Seed longevity

Longevity holds a fascination for all, and with respect to seed longevity, an early manifestation of this interest involved the investigation of the germination of seeds from old herbarium material or from seed stores. The material had been allowed to age at room temperatures under dry conditions which had not been specifically selected for the maximal retention of viability. Table 8.2 summarizes the data, quoted by Barton (1965), with respect to seeds which had maintained some viability during 40 years or more of dry storage. The surviving seeds were hard-seeded, most examples occurring in the Leguminosae, followed at some distance by the Malvaceae, Euphorbiaceae and Convolvulaceae. Some seeds maintained

Table 8.2 Families whose seeds retained viability for 40 or more years during either room temperature seed storage or storage on a herbarium sheet. Data from results of three investigators who surveyed about 300 species (Barton, 1965*b*).

Family	No of genera with viable seeds after 40 or more years of storage
Dicotyledons	
Tiliaceae	1
Malvaceae	10
Leguminosae	33
Euphorbiaceae	3
Convolvulaceae	3
Solanaceae	1
Labiatae	1
Compositae	1
Monocotyledons	
Cannaceae	1
Iridaceae	1

an ability to germinate after more than 100 years.

Dr W.J. Beal set up one of the first experiments in which the survival of weed seeds in the soil was investigated. He described the experiment in his own words (Beal, 1905) as follows:

In the autumn of 1879 I began the following experiments, with the view of learning something more in regard to the length of time the seeds of some of our most common plants would remain dormant in the soil and yet germinate when exposed to favorable conditions. I selected fifty freshly grown seeds from each of twenty-three different kinds of plants. Twenty such lots were prepared with the view of testing them at different times in the future. Each lot or set of seeds was well mixed in moderately moist sand, just as it was taken from three feet below the surface, where the land had never been plowed. The seeds of each set were well mixed with the sand and placed in a pint bottle, the bottle being filled and left uncorked, and placed with the mouth slanting downward so that water could not accumulate about the seeds. These bottles were buried on a sandy knoll in a row running east and west, and placed fifteen paces north-west from the west end of the big stone set up by the class of 1873. A bowlder stone barely even with the surface soil was set at each end of the row of bottles, which were buried about twenty inches below the surface of the ground. I should make an exception in the case of the corns, which were placed in the soil near the bottles, and not inside bottles.

Beal carried out germination tests every five years, and subsequently his successors took over the experiment and examined a bottle every 10 years, the 100-year sample being dug up in April 1980, leaving six more bottles for future tests (Kivilaan and Bandurski, 1981). The seedlings which developed were allowed to grow so as to identify them. Beal was not satisfied:

In all tests of the seeds buried in bottles, the results have been indefinite and far from satisfactory. I mean by this that I have never felt certain that I had induced all the sound seeds to germinate. I moisten the sand containing the seeds, and forthwith a goodly number germinate, and then they come slowly straggling along. I dry the soil and wait a few days, and after moistening, in a few days more seeds germinate. Why was I unable to induce them to start when treated to various degrees of temperature and moisture for seven months?

He had reason to believe that the dormancy of some viable seeds remained unbroken, and that consequently the viability of these seeds was overlooked. Although we know more tricks for breaking dormancy now, this problem remains. If such a viability experiment were set up today, very much larger numbers of each seed would be included in each sample, with the objective of improving sensitivity and permitting satisfactory statistical analysis of the data. The results for the first 100 years of this experiment are summarized in Table 8.3. The 1980 test was of interest, in that 29 seeds

Table 8.3 The viability period of buried seed in the Beal buried-seed experiments in which 50 seeds of each of 20 common species were buried in sand in pint bottles (*plus three species of tree seeds outside of the bottles*) in Michigan in 1879 (Kivilaan and Bandurski, 1981).

Viability in years	Species
< 5	*Agrostemma githago*
	Bromus secalinus
	Erechtites hieracifolia
	Euphorbia maculata
	Juglans nigra
	Quercus rubra
	Thuja occidentalis
5–10	*Trifolium repens*
30–35	*Setaria glauca*
	Stellaria media
35–40	*Capsella bursa-pastoris*
> 40	*Amaranthus retroflexus*
	Ambrosia artemesifolia
	Lepidium virginicum
	Plantago major
	Portulaca oleracea
50–60	*Brassica nigra*
	Polygonum hydropiper
80–90	*Oenothera biennis*
	Rumex crispus
> 100	*Malva rotundifolia*
	Verbascum thapsus[1]
	Verbasum blattaria[1]

[1] Both species have been identified as germinated seedlings, *V. thapsus* from 1884 to 1914 and as a single seedling in 1980, and *V. blattaria* from 1930 to 1980.

germinated, of which 6 seedlings were defective and died early before identification was possible. There was one seedling of *Malva rotundifolia*, one of *Verbascum thapsus* and 21 of *V. blattaria*.

Since Beal set up his experiment, many other studies have been conducted, and some of them are discussed with reference to the seed bank in the soil, in Chapter 7. However, mention must be made of evidence relating to the oldest known viable seeds, which were of Indian lotus (*Nelumbo nucifera*) found in a naturally drained lake bed in Manchuria, and for which a radiocarbon age of 1040 ± 210 years was obtained for seeds of the same seed lot (Barton, 1965*b*). Barton also quotes circumstantial evidence for even older viable lotus seeds—four viable seeds were discovered in the remains of a 3000-year-old canoe buried 6 m deep at a Japanese site. There is also some circumstantial evidence that seeds have maintained viability for several thousand years in the Arctic permafrost.

CHAPTER NINE

SEED TECHNOLOGY

Seed technology has been an essential foundation of all the major human civilizations and is a necessary part of agriculture, horticulture and forestry. For convenience, or for the maintenance of genotypes that would be lost in meiotic segregation and recombination, many plants (crops, ornamentals, fruits, etc.) are reproduced vegetatively by traditional means, and increasingly by the application of plant biotechnology through the multiplication of plant cells in culture followed by callus formation and the development of plantlets. Since plantlet development involves embryoid formation, a new area of biotechnology is the conversion of embryoids into artificial seeds.

However, for the present it is necessary to concentrate attention on the majority of agricultural and horticultural crops which are grown as annuals and are reproduced by seed. The developed world has a substantial seed industry concerned on one hand with the production of improved varieties and their rapid multiplication and sale, and on the other hand, with a fairly constant market based on growers' requirement for seed, purchased each year from seedsmen. Carver (1980) quotes an estimate that 602000 tonnes of wheat, barley and oats were sown in the UK in the autumn and spring of 1978, of which about 70% (420000 tonnes) was provided by the seed trade, the balance being home-saved seed. Substantial portions of agriculture and horticulture are devoted to the specialist requirements of seed production.

Seed technology is somewhat steeped in tradition and is largely based on logic and common sense. The breadth of the topic and its current practice are well covered by Thomson (1979) and are displayed diagrammatically in Table 9.1. The present chapter lays emphasis on those areas of seed technology where the plant sciences have made substantial contributions. In one sense, seed technology can be thought of as an intermediate technology which is capable of fairly straightforward adaptation to the needs of the developing world as well as those of the developed world. However, in the developed world, seed technology has become an almost wholly mechanized process dependent on high technology, whereas at farm level in the developing world, seed technology is largely a manual process

Table 9.1 Some aspects of seed technology

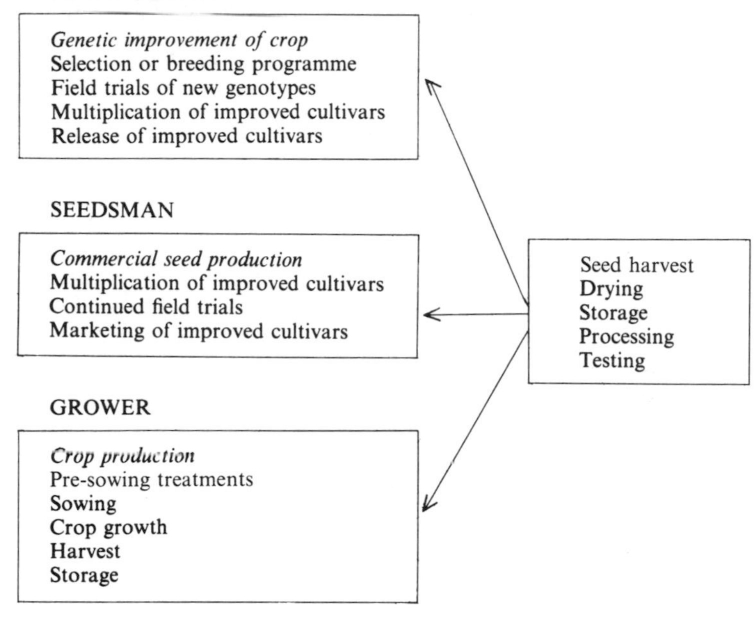

with perhaps some dependence on animal-driven machinery. Comment will be made in this chapter, when the strategies and tactics of developed and developing worlds differ.

Seed quality and testing

These are of basic importance to seed technology. Two parameters of seed quality, those of viability and vigour, were considered in Chapter 8. The third main parameter is that of seed purity. A batch of seed should be provided with a figure for its analytical purity, which is the percentage by weight of intact seeds of the species named on the label, in the whole sample. A figure for analytical purity can be determined by analysis of a small sample of seed in the seed-testing laboratory. The nature of the impurities may then be assessed: inert material may include broken seeds, chaff, other parts of the plant, and soil, while seeds of other crops and weeds are harmful. Impurities which differ from the required seeds in such characters as size, shape, density or surface properties can be reduced or eliminated by appropriate cleaning operations with mechanical processing equipment. For the determination of the content of undesirable seeds, such as those of other crops or weeds, it is necessary to determine species purity by the analysis of larger seed samples. The provision of a few extra weed seeds of a species already present on an agricultural holding would not cause serious harm, but some noxious weeds, such as wild oat (*Avena fatua*) or black grass (*Alopecurus myosuroides*), which are difficult to eradicate but are still not universally distributed, are most undesirable. Therefore, such seeds tend to be specifically mentioned in seed regulations, and seedsmen need to certify that their seeds are free from such contamination.

The examination and testing of seeds by seedsmen and government agencies is a step on the way from developing-country status to developed status. The international trade in seeds brought about the need for consistent procedures, to be used by national seed testing agencies, which evolved through the formation of the International Seed Testing Association (ISTA) in 1924. By discussion and comparative testing, the staff of the official seed testing laboratories agreed on the definitions of seed germination and purity and on the agreed methods for seed sampling and testing which ISTA published as rules. ISTA meets regularly so as to update the rules and improve seed testing procedures (see for example Table 8.1).

Not all aspects of seed purity can be ascertained by laboratory tests. Although it should be possible to identify the species present in seed samples, it is not usually possible to distinguish different cultivars visually and to determine their purity. Identification of cultivars is possible after the extraction of the seed proteins and their separation by electrophoresis. Furthermore, laboratory testing of seed does not always provide a reliable guide to the presence of disease in the seed sample. Seed-borne diseases involve viruses, mycoplasmas, bacteria, fungi and nematodes. The necessary information can be provided in a field test in which a substantial block of the seed sample is sown so that the identity and purity of the cultivar can be confirmed at the optimal time, and the incidence of disease can be ascertained. Effectively this objective is largely attained in the developed world through the inspection and certification of the seed crops grown for seed production. In urgent cases, or when quarantine is necessary, the purity of seed lots and of seed-borne diseases can be determined by germination and growth of seeds in a plot in a glasshouse.

The moisture content of a seed sample is an important factor affecting the maintenance of seed viability during storage. (see p. 101). The keeping quality of seeds can be improved after harvest by adjustment of the moisture content, usually by drying in a drying plant. Seed size is another parameter of seed quality. For some crops, large seeds may be desirable in that they tend to germinate more vigorously and to give larger seedlings whose size may carry over to larger yield. More important is uniformity of size, which favours mechanical seed handling and uniformity of germination and growth. Uniformity of germination and growth is an important property of seeds used in experimental work in the plant sciences, and therefore plant physiologists go to considerable trouble to obtain seed samples of uniform size (usually by sieving) and quality (by inspection and sorting).

It is important that a single batch of seeds should be uniform in quality, or in other words be thoroughly blended so that the contents of each bag are the same. This level of uniformity is rarely attained, and therefore in sampling for seed tests the samples must represent material taken from the whole seed batch. ISTA has formulated precise rules for the sampling of seed batches so that the tests are carried out with a working sample are accurately representative of the whole batch (ISTA, 1985).

Commercial seedsmen in the developed world are in a very competitive market, a competition which extends to the cultivars offered, to price, to seed pretreatments offered (see next section) and to quality levels well in excess of the minimum legally enforced standards. Vegetable seeds in particular are subject to much competitive selling of special seeds at special prices, suitable for precision drilling and claimed to produce uniform germination and seedling establishment.

Seed and crop production

Seed germination and establishment

The requirements are essentially similar for both seed and crop production. Satisfactory germination and seedling establishment is normally assisted by the use of seeds treated with appropriate fungicides and insecticides. Insecticide dressing would be provided to order because the insects are present in the environment and not expected to be introduced with the seeds. Similarly, some fungicidal seed dressings are provided to order when the land is suspected to be infected; others are provided routinely, except when requested otherwise, for common fungal diseases, and sometimes a fungicidal seed dressing is used to minimize the effects of an infection contained in the seed stock.

The use of agrochemicals as seed dressings has led to considerable concern about their toxicity and their persistence in the environment (Carson, 1963; Mellanby, 1967). It should be realized that seed treatments permit the application of these substances in the minimum effective amounts. Nevertheless, the general disrepute, amongst laymen, of agrochemicals depends to some extent on some spectacular episodes of mortality, between 1956 and 1961, of seed-eating birds and of predators higher in the food chain. After intensive investigation, it was concluded that the use of dieldrin (an organo-chlorine insecticide) as a seed dressing for spring-sown crops was responsible. In the developed world there are now either legislative bans or voluntary codes of restrictions relating to the use of certain pesticides, particularly organo-chlorine and organo-phosphorus compounds. There are also fears about the use of organo-mercury fungicides, the danger to wildlife, and their contribution to the build-up of mercury in the environment. Non-mercurial fungicides are therefore recommended as routine seed dressings, and it is suggested that the use of the highly effective organo-mercury compounds might be restricted to the first generations in the seed multiplication process of crop breeders and seed growers.

As the cost of seed is a substantial financial input into the costs of crop production, it is important that a uniform and proper crop density be obtained at the minimum seed rate, i.e. that maximum germination of the sown seed be obtained. This may be achieved by the use of seed with high vigour (see p. 98), by proper attention to the preparation of the seed bed and the sowing procedure, and by sowing when the soil temperature exceeds a minimum base value. Most temperate crops show little germination and growth below 5 °C, whereas warm temperate crops such as maize or *Phaseolus* spp. require minimum temperatures of 10–15 °C. Attempts have been made to speed up the rate of germination and improve the uniformity of emergence, for example, by allowing partial imbibition insufficient to allow the completion of germination, after which the seed is allowed to air dry and is sown in the normal way. Alternatively, the germination process can be taken further so that radicle emergence commences, in which case damage during sowing is minimized by the process of 'fluid drilling' where the seeds are protected by their suspension in a viscous gel. Neither of these techniques has found appreciable application for agricultural crops, but they have been used with vegetable crops. Other pretreatments may involve the reduction of seed size (by abrading or rubbing) until it is suitable for direct precision drilling (e.g. *Beta vulgaris*) or for pelleting with a small quantity of inert material so as to give standard-size globular pellets large enough for precision drilling. Pelleting is also a convenient way of handling very small seeds such as those of *Antirrhinum* and *Petunia*. Some chapters and Appendix I of Heydecker (1973) provide further reading.

Seed production

The prime objective of the seed grower is to produce a clean high-quality crop, free from disease, weeds, and other crops. As discussed above, high seed vigour is an essential parameter of seed quality, although there is little advice available about the crop management necessary for optimal seed production. A high yield is liable to be counterproductive with respect to seed quality, as the high seed rates (sowing) necessary for the achievement of high yields will cause problems with crop inspection, as well as encouraging the formation of smaller seeds with less vigorous germination. Seed production is considered at some length in Hebblethwaite (1980).

As wheat, barley and oats tend to be naturally inbreeding through self-fertilization, the grower needs to take only modest precautions in the form of maintaining a suitable distance from another variety and excluding the outermost rows of the plot from the seed harvest. The outer rows are liable to pick up wind-borne foreign pollen. One of the requirements for a new variety of these species is that it be genetically homozygous and therefore true-breeding and uniform in its characters.

However, many crops, such as maize, many members of the Cruciferae, beet, white clover, *Vicia faba* and some pasture grasses, are cross-fertilized and therefore outbreeding. They tend to show a much lower degree of uniformity as a result of the heterozygosity. It is therefore of much greater importance to have the seed plots at as great a distance as possible from plots of other varieties, and to have a sufficient time-gap (perhaps six years) between the growth of one variety and the growth of another (so that weeding of plants of the first variety should be completed).

In outbreeding plants, improved yields (resulting from hybrid vigour) and more uniform plants with uniform ripening can be achieved by hybridization, the immediate product of the hybridization in the form of F1 seed being used. As F1 progeny do not breed true, the farmer or grower who uses F1 seed cannot save seed from his crop but will have to obtain fresh supplies each growing season.

In some cases quite complex hybridizations are achieved by the breeders and growers. Amongst the achievements of breeders is the production of monogerm beet seeds, which permit the use of precision drilling and eliminate the need for hand singling in crops which previously were grown from seeds borne in clusters of two or more. A striking success has been the production of single-cross F1 *Zea mays* hybrids in the United States (Curtis, 1980). This depended on the male sterility of the female parent in the cross, since its own pollen failed to grow on the stigma, only that of the pollinator brought about fertilization. In 1970 approximately 80–85% of the hybrid seed crop in the United States was produced on male-sterile female parents in which the male sterility was cytoplasmically inherited, actually in the mitochondrial genome, all of the commercial lines possessing the T (Texas) cytoplasm. In 1970, plants which possessed the T cytoplasm, both parents and hybrids, proved to be susceptible to a new race of Southern corn leaf blight (*Helminthosporum maydis*), and as the weather favoured the spread of the disease, the overall yield fell to a national average of 4500 kg ha^{-1} in comparison to the average of 5200 kg ha^{-1} over the two previous years. The loss for many growers was disastrous. However, the seed industry responded rapidly. Plants with the so-called 'normal' cytoplasm, actually a range of cytoplasms (effectively mitochondrial and plastid genomes) proved to be resistant to race T, and the industry ceased to use T cytoplasm. Seed production was contracted out to warmer parts of the USA and to the southern hemisphere for the 1970–71 winter. Sufficient resistant seed was available for use in high-risk areas in 1971, the susceptible seed was used in low-risk areas and in weather conditions less favourable to the spread of the disease, and the 1971 crop gave the then record national yield of 5530 kg ha^{-1} in the USA.

This proved to be a sharp lesson to breeders of the potential vulnerability of a dependence on a single genotype or variety of a crop, and especially of the importance of the mitochondrial and chloroplast genomes which do undergo mutation and evolution but which do not participate in meiotic segregation and recombination. Before male sterility of the T cytoplasm in *Zea mays* was exploited, the female parents in the seed production fields were emasculated (detasselled) by hand. After the discontinuation of the use of T cytoplasm, mechanical detasselling has been developed, although it remains necessary to finish off by hand.

Male sterility continues to have important applications in hybridization in many crops, however. For crops such as tomato and for many ornamentals, the value of F1 hybrid seed is high enough to justify manual emasculation and pollination.

Harvest, post-harvest treatments and storage

Seed quality can be seriously damaged by inadequacies in these procedures, which may result in varietal contamination, loss of viability and vigour, or infestation by insects or micro-organisms. Normally the seed crop will be combine-harvested and then dried in a grain drier to a moisture content which is optimal for storage. The recommendations for maximum drying temperatures in different developed countries show a wide range of variation, and this lack of precision underlines the danger of the loss of viability and vigour from this cause. In order to avoid seed loss from dehiscence in the head, it is necessary to harvest seed before it has fully dried out in the field. Grass-seed growers may be able to obtain the best seed by cutting the grass in swathes and allowing the seeds to dry out in the windrow before harvesting with a combine. The principles of storage (see Chapter 8) should be followed with due consideration for the problems presented by large-scale storage and by difficult climates such as the wet tropics (Thomson, 1979).

Genotype preservation in seed banks

Most of the plants grown in botanic gardens are grown from seed, and therefore the collection of seed has been an essential part of the establishment and maintenance of a living plant collection. Seed surplus to the requirements of a garden is usually made available for exchange or distribution, and gardens have traditionally issued an annual list of seed. Such activity is labour-intensive, and normally surplus seed was discarded after the seed harvest of the following season. A more systematic approach to seed storage commenced about 20 years ago, led by national botanic gardens and plant research organizations. There were several reasons for this development; for example, it made economic sense to reduce the labour-intensive seed collection by storage of seeds for several years, but the main objective quickly became that of the preservation of genetic material of both wild and cultivated plants which, for the latter, involved the preservation of old and unfashionable genotypes. Plant breeders recognized the need to hold a wide range of genetic material, and conservationists recognized the role of the seed bank in the maintenance of endangered species. However, at the current rate of destruction of natural habitats, especially of tropical rain forest, species are being lost without being recognized, catalogued or ever included in a seed bank.

A seed bank, when efficiently run, is a long-term experimental programme. It is usually the function of the bank to clean up the seed material to a state suitable for storage and germination testing. The germination requirements of each species need to be determined so that the seed may be tested for viability at intervals as well as being eventually used to grow on. In the Royal Botanic Gardens, Kew, Seed Unit seeds are normally allowed to dry for about six months after harvest, before being tested (Thompson and Brown, 1972). Their standard test is carried out at six different temperatures (6, 11, 16, 21, 26 and 31 °C) in incubators in which 12 h illumination by a 30 W fluorescent tube is provided each day. If good germination occurs in response to one or more of the treatments, no further testing is required. Dormant seeds are subjected to a number of the dormancy-breaking treatments discussed in Chapter 6, e.g. chilling, fluctuating temperatures, application of gibberellins, and scarification. The objective of these tests is to determine conditions which permit at least 75% germination of a sample, so that seeds can be placed in storage on the basis that their viability can be monitored in subsequent years by the use of an appropriate set of conditions. The preliminary germination testing programme may consume a large number of seeds, but once germination conditions are established, it is possible to use the seeds in the most economical way.

At the Kew Seed Unit, most of the seed is stored in a cold store at − 10 °C, with a supplementary store at 2–4 °C for those seeds which react adversely to temperatures below 0 °C. The requirement for ready access precludes storage in hermetically sealed containers or the provision of specific gaseous atmospheres other than air. The seeds are stored either in glass bottles with bakelite screw-on caps or, for larger seeds, in self-sealing plastic bags. The main objective of the Kew Seed Unit is to maintain the viability of most of the species stored for as long as possible, which in practical terms was defined as a minimum of a 15–20 year period for 80% of the species. The importance of record keeping and organization is obvious. When the viability of a seed falls to a danger point, some of the stock is germinated and grown up for seed production.

Seed technology and plant breeding

Available evidence suggests that the domestication of the earliest crops commenced about 8000–9000 years ago, since when they have been subjected to selection and improvement, especially during the last 200 years. Some crops are more recent; for example, sugar beet dates back to the 18th century, and today new crops are being developed or sought for the production of oil or energy, or for food production in the drought-stricken areas of the world.

In the present century the selection of improved crop varieties has developed a firm scientific base, as has the production of new varieties through hybridization and selection programmes, although plant breeders tend to claim that their craft is as much an art as a science. The main objective of the plant breeder has been that of increased agricultural productivity, which in the developed world is part of the growth of material prosperity, while in the developing world it is a matter of life or death. In the developed world, improved productivity has led to agricultural surpluses which are regarded as something of a scandal, as indeed they are when food stands in store in half of the world, while the other half goes to bed hungry and where death from malnutrition may be acute and spectacular in times of famine, or merely the chronic situation of precarious survival. Agricultural surplus is the margin which protects an acceptable life-style, but political and economic action are required for the management of food surplus, with excess agricultural capacity being directed towards other existing or forthcoming shortages in materials or energy. In recent years increased agricultural productivity in Western Europe has more than made up for the apparently inevitable loss of prime agricultural land to other uses.

Plant breeders also have many other objectives, such as the modification of crops for use in unfavourable environments, for there are problems with the shortness of the growing season, with frost or low temperatures during the growing season, and with drought and salinity. Experience has shown that advances by plant breeders with respect to the incorporation of disease resistance into crop varieties may not always match the speed of natural evolution of pathogenicity by fungal pathogens, and that occasional severe setbacks occur (see p. 116). When the main product of a crop is its seed, the plant breeders' main objectives concern yield and seed quality. Nutritional value and suitability for the intended purpose, for example in such cases as malting barley or bread wheat, are paramount. An important contribution to crop yield is the ability of the seeds to germinate uniformly and vigorously, possibly under conditions rather unfavourable to germination.

Although plant breeding is costly in time and resources, it is an essential component of both the maintenance and improvement of agricultural productivity. The time-scale involved in a breeding programme can be judged from the requirements for the development of a new wheat or barley variety through the hybridization of two or more varieties which possess desirable characters. Each parent might be endowed with a large number of such characters, or the hybridization might be intended to insert disease-resistance genes from a primitive variety into a high-yielding variety. Normally the hybridizations would be made by hand with glasshouse-grown material. The F1 hybrid seed would be harvested and grown in rows outside, and the process would be repeated for F2 seed. In the F2 generation it would be possible to identify plants with apparently promising recombinations of characters, and the ears from such plants would be collected and kept for further testing. The F2 seed not selected would be harvested, bulked and sown again in the hope that desirable recombinants would become evident. During the selection process, appropriate stresses, against which resistance was desired, would be applied so that resistant plants could be identified. Such stresses might include drought or exposure to disease. Ears from selected plants would be numbered and their seeds sown in numbered sub-plots for further stress and selection.

As barley and wheat are natural inbreeders, little contamination by foreign genes takes place. In the course of several generations a typical programme will pare down perhaps 10000 plants originally selected to some which seem suitable for more extensive trials. Up to this stage, some six generations of plants will have been produced. Should a new and marketable cultivar arise from this procedure, the necessary field trial and seed multiplication procedures would require the production of about 13 generations of seed after the initial hybridizations. This need not take 13 growing seasons, as more than one crop can be obtained in a year by using glasshouses, or by growing crops alternately in northern and southern hemispheres. Crop improvement is covered by Simmonds (1979).

High hopes are held for the plant breeding possibilities of recombinant DNA technology, by which genetic modifications as small as a single base pair in a single gene might be made, or in which massive multigene transfer might be made, perhaps with no restrictions on the evolutionary distance between donor and recipient. At present, caution is necessary because of the limits of our knowledge of the plant sciences in such areas as biochemistry, development, molecular biology and molecular genetics, on one hand, and on the other the limitations in plant biotechnology with respect to cell culture and DNA transfer.

Vernalization

In Britain, most wheat and barley is autumn-sown. Winter varieties are sown in September or October as soon as the land is moist enough to be drilled directly or ploughed and brought up to the appropriate level of tilth for a seed bed. The seed germinates promptly, grows slowly during autumn and winter, and quite early in spring achieves complete ground cover, maximum growth rate and a good early crop. Modern cereal varieties are resistant to frost and survive winter well, provided that they escape flooding. In 1918, Gassner discovered that, when winter rye was sown in the spring, flowering occurred either if the seeds were allowed to germinate at 1–2 °C before being planted out of doors, or if the seedlings received sufficient chilling out of doors. This discovery proved to be of considerable importance in the USSR where in many regions the winters are too severe for the survival of high-yielding autumn-sown cereals. However, it was found that if partly imbibed winter wheat seeds were buried in the snow for a time prior to spring sowing, the seedlings developed the ability to flower in the same season. This process became known as vernalization, a term which came to describe the flowering-inducing chilling treatment which could be applied to plant material of any age. Vernalization is not used to describe the breaking of seed dormancy by chilling treatments.

Seed technology in the developing world

In what is now the developed world, the application of seed technology has grown fairly steadily, and at times quite rapidly, over the past 200 years. Much of it may be described as intermediate technology, suitable for direct transfer into the developing world. The extant scientific literature of seed research provides a strong base for the resolution of problems in the developing world, though there are difficulties with respect to the identification and location of sources in what is a fairly diffuse literature, there being relatively few specific seed journals. Training in the resolution of new problems, on a basis of published research on analogous problems, is essential. In the one project of this type undertaken in the author's laboratory, that on jute seed viability described in Chapter 8, the experimental approach was based on published accounts of viability investigations of other species (both tropical and temperate).

The ultimate solution to a problem in seed technology, whether in the developed or developing world, must depend on field work and trials in the appropriate locality. For example, prior to adoption, cultivars must be subjected to field trials in the areas for which they are intended. Application of improved seed technology offers advantages throughout the world, although perhaps the most spectacular advances can be made in the developing world, where there is greater opportunity for seed stock improvement through breeding programmes followed by improved storage and seed technology up to sowing. The British Tropical Development and Research Institute provides a good example of an aid programme aimed at the prevention of damage, from all sources, to seeds and other materials in storage.

CHAPTER TEN

AN EXPERIMENTAL APPROACH TO SEED BIOLOGY

The experimental approach has been stressed in most of the preceding chapters, commencing with the description of the germination test in Chapter 1. In most cases sufficient information can be found in the text, or at least in the references provided to enable similar experiments to be carried out. For example, Chapters 8 and 9 contain guidance and literature references for the investigation of the maintenance of viability and for seed testing, particularly for cultivated plants.

The approach described here has been used extensively in practical classes, mainly for first-year biology undergraduates and also for post-graduate students in plant sciences, and there seems no reason why a similar approach could not be used in schools. Our main objective has been to study the seed biology of members of the British flora. The experimental procedures are very simple and the results almost invariably provide some sort of intellectual stimulus and interest, and always produce the challenge to produce a plan for further experimentation. Since knowledge of the seed biology of the natural flora of the British Isles, or indeed of any area of the world, is rather fragmentary, the results of student investigations may be used as the basis of a research programme. In this laboratory we have done this by collecting and collating student results from their class reports. Such results must be regarded with a degree of scepticism, for although most of our students are hard-working, serious-minded and careful, some of them are also ingenious in devising new wrong ways to conduct experiments (in all seriousness). To overcome these shortcomings in human nature, we vigilantly oversee practical classes and endeavour to see that each seed sample is investigated by more than one student or student-pair. Also, each species is tested in two or more different years to reduce the effects of investigator error and year-to-year variation in seed behaviour, perhaps resulting from variation in the weather pattern. If serious and unexplained discrepancies are evident, the investigations are continued until they are resolved. We aim to examine seed from different locations to test for genetic variability in the seed biology of the species.

Procedure for investigating the seed biology of the natural flora of a temperate zone

Seeds are collected in the field at appropriate times and brought back to the laboratory for investigation. It must be noted that a complete laboratory investigation of the germination, and if present the dormancy, of a single species is an open-ended commitment. Consequently, in a fairly compact investigation it is possible to determine only the main biological features shown by the seed, although these features should help in at least the beginning of an understanding of the ecophysiology of the seed.

Seed collection

Normally the seed should be investigated in the form in which it falls from the parent plant, i.e. either as the naked seed or as a fruit or part of a fruit. For any species this may entail the possibility of many and frequent visits to its location, so that the seed can be collected at the time of natural dehiscence from the parent plant. This will be at some time after the cessation of the nutrient supply from parent plant to developing seed. In those species in which seeds dehisce over a period of weeks or months, several collections of seed could be taken, for comparison, during the season. The time of collection is the time to make estimates of the reproductive capacity of the plant in terms of the number of seeds produced by an individual plant, thus following the line of work carried out by Salisbury (1942).

The method of harvesting the seeds depends on their nature, but is often most convenient to take the complete fruit or the complete fruiting inflorescence. The material is placed in paper bags or large paper envelopes, a label is written, and they are transported back to the laboratory fairly promptly. The material is allowed to air-dry for two or three days, either in the envelopes or spread on a tray. During this time the material is sorted out, the seeds are freed from the residual plant material, and an attempt is made to free the seeds from animal life, especially from likely animal predators.

Seed treatments

Newly harvested batches of seeds should be divided into three approximately equal samples, of which one should be subjected to a set of germination tests without delay. The remaining batches are given the storage treatments which have been found to be the most effective in breaking the dormancy of species native to Britain, namely dry storage and chilling. For dry storage the sample is placed in a paper envelope on a shelf in the laboratory for 3–6 months, where it is subjected to ambient temperatures. For chilling, the seeds are mixed with about 10 times their volume of moist sand in a stoppered tube or a screw-top jar which contains at least half its volume as air. The vessel is placed in a refrigerated incubator, refrigerator or cold room at about 5 °C for a minimum period of three months.

When the seeds were used as part of a practical class programme, care was taken to see that all of the seeds had received a minimum of three months' dry-storage by the commencement of classes. If seed has been picked early in the previous growing season, a rather longer dry-storage was experienced. Since the practical classes were scheduled to start in mid-February, the chilling treatments usually start in mid-November so as to provide a minimum of three months' chilling, prior to which the seeds were held in dry storage. Thus the chilling treatment occurred at approximately the same time as chilling in the natural environment.

Since the students would not normally be present at harvest time, they would not be able to test the germination of newly-harvested seed immediately after collection. Consequently, the batches of newly-harvested seed are stored in sealed bottles in the − 20 °C deep-freeze so that their properties in the newly-harvested condition can be preserved until the commencement of the classes, when parallel experiments with 'newly-harvested' dry-stored and chilled seeds can be carried out.

Germination tests

The standard procedure for germination tests, as described in Chapter 1 (pp. 3–5) is followed, samples from each of the treatments (newly-harvested, dry-stored and chilled seeds) being tested at 5 °C, 10 °C, 15 °C and 20 °C. In these tests the seeds are manipulated and counted daily in daylight or under artificial illumination, but are incubated in standard unilluminated incubators. The seeds receive some illumination during testing, which may or may not be sufficient to affect their germination behaviour. Therefore, two additional tests are required to test the response of the seeds to light. For convenience, these tests are carried out at 20 °C. In one treatment the germination is carried out under continuous illumination of 33 $\mu E.m^{-2}.s^{-1}$ in a growth cabinet or illuminated incubator. In the other treatment, the germination test is conducted in total darkness in petri dishes covered with aluminium foil in a dark incubator. To avoid irradiation with biologically active light when germination is recorded, several replicate 'total darkness' samples are used, a fresh one being examined at each appropriate interval of the total darkness test, and then discarded.

I regard it as logical to allow the tests to run for 28 days, provided that 100% germination is not recorded in a shorter time. At the end of this time there are three distinct classes of result:

(i) High germination irrespective of treatment, temperature or illumination
(ii) No germination at all
(iii) Different amounts of germination for the different treatments and temperatures.

The occurrence of a high germination in all treatments shows an apparent lack of dormancy in that seed. The results should establish the optimum temperature for germination. It may be worthwhile to investigate the longevity of the seed, to determine whether dormancy may develop on storage or whether dormancy may be induced by such treatments as far-red irradiation, exposure to ethylene, burial in soil, etc.

A seed sample which fails to give any germination under any treatment shows either deep dormancy or low viability. Some indication about the viability may be obtained by vital staining of the seed (see p. 100) together with a microscopic examination to search for the presence and state of development of the embryo. Should the seed sample appear to have a reasonable number of viable embryos, the germination tests should be allowed to continue beyond 28 days until germination occurs. In addition, the range of treatments listed in the next section should be considered.

When one or more of the tests shows statistically significant levels of germination and when there are clear differences between the treatments, there should be the beginning of an understanding of the dormancy mechanisms and the dormancy-breaking mechanisms involved. In this case an extension of the experiments already performed may be required, experiments may need to be repeated in order to obtain an improved statistical analysis, and some of the treatments listed in the next section may be appropriate.

It should be noted that seeds of some species will germinate in the total darkness of the 5 °C chilling treatment, and that when this happens, all of the seeds may germinate. In this event, the resultant seedlings should be separated from any ungerminated seeds, normally under a low-power binocular microscope, and a determination of percentage germination should be made. If sufficient ungerminated seeds are present, these should be used in the standard germination tests. If high germination occurs in the chilling treatment, a similar result would be expected for newly-harvested seeds when tested at 5 °C.

Further investigations into breaking of seed dormancy

(i) *Removal from the fruit*. Those seeds which disperse naturally as the whole fruit or a section of a fruit can be removed from the fruit after the freezing, chilling and dry storage treatments, and re-subjected to the standard range of germination tests described in the previous section.

(ii) *Isolation of the embryo*. The occurrence of embryo dormancy can be tested by dissection of the embryos from seeds which had been subjected to the three treatments, followed by the standard germination tests.

(iii) *Scarification*. Dormancy can be broken by merely rupturing the seed coverings. This may be done with a scalpel or mounted needle. Alternatively, concentrated H_2SO_4 can be applied with great care.

(iv) *Treatment with fluctuating conditions*. For some seeds exposure to fluctuating conditions of temperature or illumination is an obligatory requirement for germination. For many seeds, fluctuating conditions result in optimum germination. Fluctuating conditions can be provided on a diurnal basis in a controlled environment cabinet or an incubator, for example, a 12 h day at 16 °C might alternate with a 12 h night at 6 °C. Long days of 16 h and short days of 8 h are alternatives.

(v) *Treatment with chemicals*. Chemicals which penetrate to the embryo and stimulate metabolic activity are often effective in bringing about germination. Four readily available chemicals are 10^{-4} M gibberellic acid (GA_3), 10^{-4} M kinetin, 0.1 M thiourea and 0.2% KNO_3.

(vi) *Treatment with gases*. Although carbon dioxide and ethylene may be responsible for the imposition of dormancy, they also can break dormancy, either singly or together. Since they are naturally produced by living organisms, including the seeds which may respond to their accumulation, any treatment with gas needs to be accompanied by appropriate controls kept free from accumulation of the gases by the use of filter paper wicks soaked in KOH (for CO_2) and potassium permanganate (for ethylene). 250-ml conical flasks sealed by serum bottle caps make suitable germination vessels. They can be flushed with the desired gaseous atmosphere by means of two hypodermic needles, one for input and one for output.

(vii) *Leaching*. There may be no germination until prolonged leaching of the seeds has occurred. Seed which leach dark-coloured components in a petri dish may fall into this category. Leaching can be achieved by changing the water in a petri dish at daily intervals. Alternatively, an apparatus by which the seeds are continually leached by a slow flow of water might be tried. The leachate can be collected and tested on non-dormant seeds.

(viii) *Light*. If the experiments in the previous section provide evidence that light promotes or inhibits germination, a more critical appraisal of the light effects should be carried out. The determination of the light requirement in terms of duration, intensity and quality of illumination is liable to be a substantial undertaking.

Problems with infected seed samples

Seeds are normally resistant to microbial infection and it is not usually necessary to adopt aseptic techniques. However, it is possible for seed samples to be heavily contaminated with virulent pathogens, and it is quite common for moulds to set up heavy colonies on dead seeds and the dead outer coverings of viable seeds. These infections are favoured by the high humidities within the seed-test petri dishes, and under such conditions the fungal exudates can damage germinating seedlings and the mycelium can invade or overgrow seeds and seedlings. Some useful precautions are to avoid testing damaged or infected seeds, and to remove any infected seeds from the test container as soon as infection is evident. This should be done during the daily recording of germination, the number of infected germinated or ungerminated seeds so removed being recorded.

When seed samples develop heavy infections it may be necessary to surface-sterilize the seeds by immersion in a sodium hypochlorite solution (1% available chlorine) for 10 minutes (or longer), after which the seeds should be washed in sterile water. In this case it may or may not be necessary to carry out the germination tests under strictly aseptic conditions. Surface sterilization will not be effective against micro-organisms buried beneath the surface of the seed. An alternative or additional approach is to add antibiotics or fungicides such as penicillin, streptomycin, chloramphenicol or nistatin to the germination dish, though these compounds may affect the germination of the seeds.

The seed bank in the soil

Soil seed banks are discussed extensively in Chapter 7, and many ideas for class or research investigations can be developed from the literature quoted there. For class purposes we find it most convenient for students to test one litre of sieved soil in a 16 × 22 cm seed tray. The soil is watered and the tray is then kept moist and free from any further seed rain. Strictly speaking, the trays should be held in a house constructed of a fine mesh which prevents any fresh seed rain, but which allows the trays to be subjected to ambient conditions of temperature and illumination. However, when class experiments are carried out in winter it is necessary to speed up germination by means of a heated glasshouse. All the seedlings are identified, counted and recorded, and removed from the trays at weekly or longer intervals. Identification of seedlings is quite difficult, even if one is supplied with ample illustrated handbooks such as Muller (1978). Often it is necessary to grow-on some seedlings until they attain sufficient of their adult characters to permit unequivocal identification. A systematic approach in which identified seedlings, at different stages of their development, are pressed, dried and mounted on cards should provide one with a valuable reference collection.

When germination in the trays falls off appreciably, all the seedlings should be removed from each tray and the soil should be stirred thoroughly so that some of the lower layers are brought to the surface and some of the upper layers are buried. Germination is again followed. For an exhaustive survey, a sample of soil should be tested for 18 months during which it should be stirred every 3 months.

Seed biology as an open-ended research problem

Adaptations may be made to adjust these procedures to class projects or research use. For class purposes, it may be necessary to confine the recording process to one day per week. For both research and teaching, it will probably be necessary to define the procedure by adjusting the timing of the operations, by new combinations of treatments and by the introduction of novel treatments. As data accumulate, new experiments should be devised with the objective of improving understanding of the biology of the seed under investigation. As in any area of biology, confidence in one's data is required before one can proceed to the next step. It is necessary to perform the requisite controls, to repeat experiments and to use correct statistical procedures. The 95% confidence limits devised by Roberts (1963) (see Appendix) are most convenient as a basic statistical method.

My experimental protocol for the investigation of the germination characteristics of the natural vegetation of Britain is outlined in Table 10.1. An alternative procedure, employed as part of the research programme of the NERC Unit of Comparative Plant Ecology in the University of

Table 10.1 An experimental protocol for the investigation of germination and dormancy characteristics of the seeds of some British plants.

1. Collect ripe seed.
2. Identify, sort and clean seed.
3. Air-dry the seed in the open laboratory for 2 to 3 days.
4. Split the seed sample into three equal samples and pretreat as follows:
   A. Either test germination immediately as in (5) below, or place the seed in a sealed container in the − 20 °C deep freeze for later testing
   B. Store air dry seed in an unsealed container at laboratory temperature for 3 months
   C. Chill at 5 °C in moist sand for 3 months.
5. Test the germination of each of samples 4A, B and C under the following conditions:
   I 5 °C unilluminated
   II 10 °C "
   III 15 °C "
   IV 20 °C "
   V 20 °C under continuous illumination.
   VI 20 °C under total darkness.
6. Wherever appropriate, carry out germination tests under the following treatments:
   (a) After removal from the fruit
   (b) With the naked embryo
   (c) After scarification
   (d) During exposure to fluctuating conditions of light and temperature
   (e) During treatment with chemicals
   (f) During treatments with gases
   (g) After leaching
   (h) In response to illumination treatments.

Sheffield, has been applied by Grime *et al.* (1981) to a study of the germination of seeds of 403 species collected in the Sheffield region of Great Britain.

APPENDIX

Simple procedure for determining statistical significance of results of germination tests

A convenient way of determining whether the differences between treatments in germination tests are statistically significant was developed by E.H. Roberts (1963). It involves the calculation of the limits of χ^2 insignificance at the 5% probability level for a single germination test. In other words, if the value for percentage germination of another treatment lies outside these limits, the difference between it and the first test is significant. Roberts calculated the limits of insignificance as follows. On the basis of 100 seeds used in each treatment in the germination test and with a value of χ^2 of 3.84 (the value appropriate for one degree of freedom at the 5% probability level), the following quadratic equation was derived:

$$x^2 - 3.768x - 1.923xy - 3.768y + y^2 = 0$$

For each whole digit percentage value of x, the alternative solutions for y gave a higher value and a lower value. Yates' correction was applied by subtracting 0.5 from the higher value of y and by adding 0.5 to the lower value of y. For a given value of x, the alternative solutions for y give the minimum higher and maximum lower values which achieve significance. In Table A1, the values of y for each value of x are expressed as the nearest whole number which lies in the area of significance. These numbers are calculated for a sample size of 100 seeds. The use of a different sample size would require a complete recalculation of the table, of course.

Roberts pointed out that sometimes a treatment caused a slight effect which was insufficient to achieve significance in any single test. In those cases where it was important to know whether an effect, though slight, was real, a further χ^2 analysis was carried out with the sums of the germination values for the series of tests.

Table A1 The alternative percentage values (y) which are significantly different from a value (x) when using 100 seeds in each treatment. The table includes Yates' correction.

x	y	y	x	y	y	x	y	y	x	y	y
0	6	—	—	—	—	—	—	—	—	—	—
1	8	—	26	41	13	51	66	36	76	88	62
2	10	—	27	42	14	52	67	37	77	89	63
3	12	—	28	43	15	53	68	38	78	90	64
4	13	—	29	44	16	54	69	39	79	91	65
5	15	—	30	45	17	55	70	40	80	91	66
6	16	0	31	46	18	56	71	41	81	92	67
7	18	0	32	47	18	57	72	42	82	93	68
8	19	1	33	48	19	58	73	43	83	93	70
9	20	1	34	49	20	59	74	44	84	94	71
10	21	2	35	50	21	60	75	45	85	95	72
11	23	2	36	51	22	61	75	46	86	96	73
12	24	3	37	52	23	62	76	47	87	96	75
13	25	4	38	53	24	63	77	48	88	97	76
14	27	4	39	54	25	64	78	49	89	98	77
15	28	5	40	55	25	65	79	50	90	98	78
16	29	6	41	56	26	66	80	51	91	99	80
17	30	7	42	57	27	67	81	52	92	99	81
18	32	7	43	58	28	68	82	53	93	100	82
19	33	8	44	59	29	69	82	54	94	100	84
20	34	9	45	60	30	70	83	55	95	—	85
21	35	9	46	61	31	71	84	56	96	—	87
22	36	10	47	62	32	72	85	57	97	—	88
23	37	11	48	63	33	73	86	58	98	—	90
24	38	12	49	64	34	74	87	59	99	—	92
25	40	13	50	65	35	75	87	60	100	—	94

REFERENCES

Abdalla, F.H. and Roberts, E.H. (1968) Effects of temperature, moisture, and oxygen on the induction of chromosome damage in seeds of barley, broad beans, and peas during storage. *Ann. Bot. N.S.* **32**, 119–136.

Arias, I., Williams, P.M. and Bradbeer, J.W. (1976) Studies in seed dormancy. IX. The role of gibberellin biosynthesis and the release of bound gibberellin in the post-chilling accumulation of gibberellin in seeds of *Corylus avellana* L. *Planta* (*Berlin*) **131**, 135–139.

Bachelard, E.P. (1967) Role of the seed coat in dormancy of *Eucalyptus pauciflora* and *E. delegatensis* seeds. *Aust. J. Biol. Sci.* **20**, 1237–1240.

Bartley, M.R. and Frankland, B. (1982) Analysis of the dual role of phytochrome in the photoinhibition of seed germination. *Nature* (*London*) **300**, 750–752.

Barton, L.V. (1965*a*) Seed dormancy. General survey of dormancy types in seeds, and dormancy imposed by external agents. In *Encyclopedia of Plant Physiology*, ed. W. Ruhland, Vol. **XV/2**, Springer, Berlin and Heidelberg, 699–720.

Barton, L.V. (1965*b*) Longevity in seeds and in the propagules of fungi. In *Encyclopedia of Plant Physiology*, ed. W. Ruhland, Vol. **XV/2**, Springer, Berlin and Heidelberg, 1058–1085.

Baskin, J.M. and Baskin, C.C. (1984) Role of temperature in regulating timing of germination in soil seed reserves of *Lamium purpureum* L. *Weed Res.* **24**, 341–349.

Baskin, J.M. and Baskin, C.C. (1985) Does seed dormancy play a role in the germination ecology of *Rumex crispus? Weed Sci.* **33**, 340–343.

Beadle, N.C.W. (1940) Soil temperatures during forest fires and their effect on the survival of vegetation. *J. Ecol.* **28**, 180–192.

Beal, W.J. (1905) The vitality of seeds. *Bot. Gaz.* **40**, 140–143.

Beevers, H. (1975) Organelles from castor bean seedlings: biochemical roles in gluconeogenesis and phospholipid biosynthesis. In *Recent Advances in the Chemistry and Biochemistry of Plant Lipids*, eds. T. Galliard and E.I. Mercer, Academic Press, London, 287–299.

Berg, R.Y. (1975) Myrmecochorous plants in Australia and their dispersal by ants. *Aust. J. Bot.* **23**, 477–508.

Berjak, P. and Villiers, T.A. (1972) Ageing in plant embryos. IV. Loss of regulatory control in aged embrvos. *New Phytol.* **71**, 1069–1074.

Bewley, J.D. and Black, M. (1978) *Physiology and Biochemistry of Seeds in Relation to Germination*. Vol. 1, *Development, Germination and Growth*, Springer, Berlin and Heidelberg.

Bewley, J.D. and Black, M. (1982) *Physiology and Biochemistry of Seeds in Relation to Germination*. Vol. 2, *Viability, Dormancy and Environmental Control*, Springer, Berlin and Heidelberg.

Borthwick, H.A. (1972) History of phytochrome. In *Phytochrome*, eds. K. Mitrakos and W. Shropshire, Academic Press, London, 3–23.

Borthwick, H.A., Hendricks, S.B., Toole, E.H. and Toole, V.K. (1954) Action of light on lettuce seed germination. *Bot. Gaz.* **115**, 205–225.

Bourque, J.E. and Wallner, S.J. (1982) Endosperm and pericarp involvement in the supercooling of imbibed lettuce seeds. *Plant Physiol.* **70**, 1571–1573.

Bradbeer, J.W. (1968) Studies in seed dormancy. IV. The role of inhibitors and gibberellin in the dormancy and germination of *Corylus avellana* L. seeds. *Planta* (*Berlin*) **78**, 266–276.

Bradbeer, J.W. (1981) Development of photosynthetic function during chloroplast bio-

genesis. In *The Biochemistry of Plants*, eds. P.K. Stumpf and E.E. Conn. Vol. 8, *Photosynthesis* eds. M.D. Hatch and N.K. Boardman, Academic Press, New York, 423–472.
Bradbeer, J.W. and Colman, B. (1967) Studies in seed dormancy. I. The metabolism of [$2^{14}C$] acetate by chilled seeds of *Corylus avellana* L. *New Phytol.* **66**, 5–15.
Bradbeer, J.W. and Pinfield, N.J. (1967) Studies in seed dormancy. III. The effects of gibberellin on dormant seeds of *Corylus avellana* L. *New Phytol.* **66**, 515–523.
Bradbeer, J.W., Arias, I.E. and Nirmala, I.S. (1978) The role of chilling in the breaking of seed dormancy in *Corylus avellana* L. *Pesticide Sci.* **9**, 184–186.
Bradstock, R.A. and Myerscough, P.J. (1981) Fire effects on seed release and the emergence and establishment of seedlings in *Banksia ericifolia* L.f. *Aust. J. Bot.* **29**, 521–531.
Brenchley, W.E. and Warington, K. (1930) The weed seed population of arable soil. 1. Numerical estimation of viable seeds and observations on their natural dormancy. *J. Ecol.* **18**, 235–272.
Briarty, L.G. (1980) Stereological analysis of cotyledon cell development in *Phaseolus*. II. The developing cotyledon. *J. exp. Bot.* **31**, 1387–1398.
Butler, W.L., Norris, K.H., Siegelman, H.W. and Hendricks, S.B. (1959) Detection, assay and preliminary purification of the pigment controlling photoresponsive development of plants. *Proc. Natl. Acad. Sci. USA* **45**, 1703–1708.
Carlquist, S. (1967) The biota of long distance dispersal. V. Plant dispersal to Pacific Islands. *Bull. Torrey Bot. Club* **94**, 129–162.
Carson, R. (1963) *Silent Spring*. Hamish Hamilton, London.
Carver, M. (1980) The production of quality cereal seed. In *Seed Production*, ed. P.D. Hebblethwaite, Butterworth, London, 295–306.
Chippindale, H.G. and Milton, W.E.J. (1934) On the viable seeds present in soil beneath pastures. *J. Ecol.* **22**, 508–531.
Côme, D. (1980/81) Problems of embryonal dormancy as exemplified by apple embryo. *Israel J. Bot.* **29**, 145–156.
Cone, J.W., Jaspers, P.A.P.M. and Kendrick, R.E. (1985) Biphasic fluence-response curves for light induced germination of *Arabidopsis thaliana* seeds. *Plant Cell Environ.* **8**, 605–612.
Corner, E.J.H. (1976) *The Seeds of Dicotyledons*. 2 vols, Cambridge University Press, Cambridge.
Corvillon, E. and Martinez-Honduvilla, C.J. (1980) Germination inhibitors in embryos and endosperms from *Pinus pinea* seeds. *Ital. J. Biochem.* **29**, 405–411.
Crocker, W. and Barton, L.V. (1953) *Physiology of Seeds*. Chronica Botanica, Waltham, Mass.
Crocker, W. and Davis, W.E. (1914) Delayed germination in seed of *Alisma plantago*. *Bot. Gaz.* **58**, 285—321.
Curtis, D.L. (1980). Some aspects of *Zea mays* L. (corn) seed production in the U.S.A. In *Seed Production*, ed. P.D. Hebblethwaite, Butterworth, London, 389–400.
Del Tredici, P. and Torrey, J.G. (1976) On the germination of seeds of *Comptonia peregrina*, the sweet fern. *Bot. Gaz.* **13**, 262–268.
Dure, L.S. III. (1975) Seed formation. *Ann. Rev. Plant Physiol.* **26**, 259–278.
Dure, L.S. III. Capdevila, A.M. and Greenway, S.C. (1979) Messenger RNA domains in the embryogenesis and germination of cotton cotyledons. In *Genome Organization and Expression in Plants*, ed. C.J. Leaver, Plenum, New York, 127–146.
Enu-Kwesi, L. and Dumbroff, E.B. (1978) Changes in abscisic acid in the embryo and covering structures of *Acer saccharum* during stratification. *Zeitschr. Pflanzenphysiol.* **86**, 371–377.
Enu-Kwesi, L. and Dumbroff, E.B. (1980) Changes in phenolic inhibitors in seeds of *Acer saccharum* during stratification. *J. exp. Bot.* **31**, 425–436.
Esashi, Y. and Leopold, A.C. (1968) Physical forces in dormancy and germination of *Xanthium* seeds. *Plant Physiol.* **43**, 871–876.
Evenari, M. (1965). Light and seed dormancy. In *Encyclopedia of Plant Physiology*, ed. W. Ruhland, Vol. **XV/2**, Springer, Berlin and Heidelberg, 804–847.
Feierabend, J. (1979) Role of cytoplasmic protein synthesis and its coordination with the

plastidic protein synthesis in the biogenesis of chloroplasts. *Ber. Dtsch. Bot. Ges.* **92**, 553–594.

Fenner, M. (1985) *Seed Ecology*. Chapman and Hall, London.

Foster, A.S. and Gifford, E.M. Jr. (1974) *Comparative Morphology of Vascular Plants*. 2nd edn., W.H. Freeman, San Francisco.

Frankland, B. and Wareing, P.F. (1966) Hormonal regulation of seed dormancy in hazel (*Corylus avellana* L.) and beech (*Fagus sylvatica* L.) *J. exp. Bot.* **17**, 596–611.

Goodwin, T.W. and Mercer, E.I. (1983) *Introduction to Plant Biochemistry*. 2nd edn, Pergamon, Oxford.

Grime, J.P. (1979) *Plant Strategies and Vegetative Processes*. John Wiley, Chichester.

Grime, J.P., Mason, G., Curtis, A.V., Rodman, J., Band, S.R., Mowforth, M.A.G., Neal, A.M. and Shaw, S. (1981) A comparative study of germination characteristics in a local flora. *J. Ecol.* **69**, 1017–1059.

Groves, J.F. (1917) Temperature and life duration of seeds. *Bot. Gaz.* **63**, 169–189.

Gulliver, R.L. and Heydecker, W. (1973) Establishment of seedlings in a changeable environment. In *Seed Ecology*, ed. W. Heydecker, Butterworth, London, 433–462.

Hamilton, D.F. and Carpenter, P.L. (1975) Regulation of seed dormancy in *Elaeagnus umbellata* by endogenous growth substances. *Can. J. Bot.* **53**, 2303–2311.

Hanna, P.J. (1984) Anatomical features of the seed coat of *Acacia kempeana* (Mueller) which relate to increased germination rate, induced by heat treatment. *New Phytol.* **96**, 23–29.

Harborne, J.B. (1982) *Introduction to Ecological Biochemistry*. Academic Press, London.

Harper, J.L. (1957) Biological flora of the British Isles, *Ranunculus acris* L., *Ranunculus repens* L., *Ranunculus bulbosus* L. *J. Ecol.*, **45**, 289–342.

Harper, J.L. (1977) *Population Biology of Plants*. Academic Press, London.

Hebblethwaite, P.D. (1980) *Seed Production*. Butterworth, London and Boston.

Hemmat, M., Zeng, G.-W. and Khan, A.A. (1985) Responses of intact and scarified curly dock (*Rumex crispus*) seeds to physical and chemical stimuli. *Weed Sci.* **33**, 658–664.

Heydecker, W. (1972) Vigour. In *Viability of Seeds*, ed. E.H. Roberts, Chapman and Hall, London, 209–252.

Heydecker, W. (ed.) (1973) *Seed Ecology*. Butterworth, London.

Hilton, J.R. (1984) The influence of light and potassium nitrate on the dormancy and germination of *Avena fatua* L. (Wild Oat) seed and its ecological significance. *New Phytol.* **96**, 31–34.

Hyde, E.O.C. (1954) The function of the hilum in some Papilionaceae in relation to the ripening of the seed and the permeability of the testa. *Ann. Bot. N.S.* **70**, 241–256.

Ikuma, H. and Thimann, K.V. (1963) The role of the seed-coats in germination of photosensitive lettuce seeds. *Plant Cell Physiol.* **4**, 169–185.

International Seed Testing Association (1985) International rules for seed testing. Rules and Annexes, 1985. *Seed Sci. and Technol.* **13**, 299–515.

Jackson, G.A.D. (1968) Hormonal control of fruit development, seed dormancy and germination with particular reference to *Rosa*. In *Plant Growth Regulators*, Society of Chemical Industry Monograph No. 31, 127–156.

Jackson, G.M. and Varriano-Marston, E. (1980) A simple autoradiographic technique for studying diffusion of water into seeds. *Plant Physiol.* **65**, 1229–1230.

Jacobs, W.P. (1962) Longevity of plant organs: internal factors controlling abscission. *Ann. Rev. Plant Physiol.* **13**, 403–436.

Jacobsen, J.V. and Knox, R.B. (1974) The proteins released by isolated barley aleurone layers before and after gibberellic acid treatment. *Planta (Berlin)* **115**, 193–206.

Jain, N.K. and Saha, J.R. (1971) Effect of storage length on seed germination in jute (*Corchorus spp.*). *Agronomy J.* **63**, 636–638.

Jarvis, B.C., Wilson, D.A. and Fowler, M.W. (1978) Growth of isolated embryonic axes from dormant seeds of hazel (*Corylus avellana* L.). *New Phytol.* **80**, 117–123.

Jones, R.L. and Stoddart, J.L. (1977) Gibberellins and seed germination. In *The Physiology and Biochemistry of Seed Dormancy and Germination*, ed. A.A. Khan, North-Holland, Amsterdam, 77–109.

Keefe, P.D. and Moore, K.G. (1981) Freeze desiccation: a second mechanism for the survival of hydrated lettuce (*Lactuca sativa* L.) seed at sub-zero temperatures. *Ann. Bot.* **47**, 635–645.

Khandakar, A.L. and Bradbeer, J.W. (1983) *Jute Seed Quality*. Bangladesh Agricultural Research Council, Dhaka.

Kirk, J.T.O. and Tilney-Bassett, R.A.E. (1978) *The Plastids*. 2nd edn., Elsevier/North Holland, Amsterdam.

Kivilaan, A. and Bandurski, R.S. (1981) The one hundred-year period for Dr Beal's seed viability experiment. *Amer. J. Bot.* **68**, 1290–1292.

Kriedemann, P. and Beevers, H. (1967) Sugar uptake and translocation in the castor bean seedling. I. Characteristics of transfer in intact and excised seedlings. *Plant Physiol.* **42**, 161–173.

Lang, A. (1965) Effects of some internal and external conditions on seed germination. In *Encyclopedia of Plant Physiology*, ed. W. Ruhland, Vol. XV/2, Springer, Berlin and Heidelberg, 848–893.

Le Page-Degivry, M.-Th. (1973) Intervention d'un inhibiteur lié dans la dormance embryonnaire de *Taxus baccata* L. *C.r. Acad. Sci. Paris, Séc. D*, **277**, 177–180.

Le Page-Degivry, M.-Th. and Bulard, C. (1979) Acide abscissique lié et dormance embryonnaire chez *Pyrus malus*. *Physiol. Plant.* **46**, 115–120.

Mayer, A.M. and Poljakoff-Mayber, A. (1982) *The Germination of Seeds*. Pergamon, Oxford.

Mellanby, K. (1967) *Pesticides and Pollution*. Collins, London and Glasgow.

Mohr, H. (1972) *Lectures on Photomorphogenesis*. Springer, Berlin, Heidelberg, New York.

Monin, J. (1967) Étude d'un inhibiteur existant chez les embryons dormants d'*Euonymus europaeus* L. *C.r. Acad, Sci. Paris, Sér. D.* **264,** 2367–2370.

Muller, F.M. (1978) *Seedlings of the North-western European Lowland. A Flora of Seedlings.* Dr. W. Junk, The Hague.

Murray, D.R. (ed.) (1984) *Seed Physiology*. Vols. 1 and 2, Academic Press Australia.

Murray, D.R. (ed.) (1987) *Seed Dispersal*, Academic Press, Sydney.

Naqvi, H.H. and Hanson, G.P. (1982) Germination and growth inhibitors in guayule (*Parthenium argentatum* Gray) chaff and their possible influence in seed dormancy. *Amer. J. Bot.* **69**, 985–989.

Nikolaeva, M.G. (1967) *Physiology of Deep Dormancy in Seeds*, Izdatel' stvo "Nauka" Leningrad, published in translation for the National Science Foundation, Washington, D.C., by the Israel Program for Scientific Translations, Jerusalem, 1969.

Nikolaeva, M.G. (1977) Factors controlling the seed dormancy pattern. In *The Physiology and Biochemistry of Seed Dormancy and Germination*, ed. A.A. Khan, North-Holland, Amsterdam, 51–74.

Öpik, H. (1965) Respiration rate, mitochondrial activity and mitochondrial structure in the cotyledons of *Phaseolus vulgaris* L. during germination. *J. exp. Bot.* **16**, 667–682.

Öpik, H. (1966) Changes in cell fine structure in the cotyledons of *Phaseolus vulgaris* L. during germination. *J. exp. Bot.* **17**, 427–439.

Öpik, H. (1968) Development of cotyledon cell structure in ripening *Phaseolus vulgaris* seeds. *J. exp. Bot.* **19**, 64–76.

Öpik, H. and Simon. E.W. (1963) Water content and respiration rate of bean cotyledons. *J. exp. Bot.* **14**, 299–310.

Osborne, D.J. (1982) Deoxyribonucleic acid integrity and repair in seed germination: the importance in viability and survival. In *The Physiology and Biochemistry of Seed Development, Dormancy and Germination*, ed. A.A. Khan, Elsevier, Amsterdam, 435–463.

Phillips, I.D.J. and Jones, R.L. (1964) Gibberellin-like activity in bleeding-sap of root systems of *Helianthus annuus* detected by a new dwarf pea epicotyl assay and other methods. *Planta (Berlin)* **63**, 269–278.

Pieniazek, J. and Grochowska, M.J. (1967) The role of the natural growth inhibitor (Abscisin II) in apple seed germination and the changes in the content of phenolic substances during stratification. *Acta Soc. Bot. Pol.* **36**, 579–587.

Porter, N.G. and Wareing, P.F. (1974) The role of the oxygen permeability of the seed coat in the dormancy of seed of *Xanthium pennsylvanicum* Wallr. *J. exp. Bot.* **25**, 583–594.

Raghavan, V. (1976) *Experimental Embryogenesis in Vascular Plants*, Academic Press, London.

Roberts, E.H. (1961*a*) Viability of cereal seed for brief and extended periods. *Ann. Bot. N.S.* **25**, 373–380.

Roberts, E.H. (1961*b*) The viability of rice seed in relation to temperature, moisture content and gaseous environment. *Ann. Bot. N.S.* **25**, 381–390.

Roberts, E.H. (1963) The effects of inorganic ions on dormancy in rice seed. *Physiol. Plant.* **16**, 732–744.

Roberts, E.H. (1964) The distribution of oxidation-reduction enzymes and the effects of respiratory inhibitors and oxidising agents on dormancy in rice seed. *Physiol. Plant.* **17**, 14–29.

Roberts, E.H. (1965) Dormancy in rice seed. IV. Varietal responses to storage and germination temperatures. *J. exp. Bot.* **16**, 341–349.

Roberts, E.H. (1972) Storage environment and the control of viability. In *Viability of Seeds*, ed. E.H. Roberts, Chapman and Hall, London, 14–58.

Roberts, E.H. and Roberts, D.L. (1972) Viability nomographs. In *Viability of Seeds*, ed. E.H. Roberts, Chapman and Hall, London, 417–423.

Roberts, E.H. and Smith, R.D. (1977) Dormancy and the pentose phosphate pathway. In *The Physiology and Biochemistry of Seed Dormancy and Germination*, ed. A.A. Khan, Elsevier, Amsterdam, 385–411.

Roos, E.E. (1982) Induced genetic changes in seed germplasm during storage. In *The Physiology and Biochemistry of Seed Development, Dormancy and Germination*, ed. A.A. Khan, Elsevier, Amsterdam, 409–434.

Ross, J.D. and Bradbeer, J.W. (1971*a*) Studies in seed dormancy. V. The content of endogenous gibberellins in seeds of *Corylus avellana* L. *Planta (Berlin)* **100**, 288–302.

Ross, J.D. and Bradbeer, J.W. (1971*b*). Studies in seed dormancy. VI. The effects of growth retardants on the gibberellin content and germination of chilled seeds of *Corylus avellana* L. *Planta (Berlin)* **100**, 303–308.

Salisbury, E.J. (1942) *The Reproductive Capacity of Plants: Studies in Quantitative Biology*. Bell, London.

Sarukhán, J. (1974) Studies on plant demography: *Ranunculus repens* L., *R. bulbosus* L. and *R. acris* L. II. Reproductive strategies and seed population dynamics. *J. Ecol.* **62**, 151–177.

Simmonds, N.W. (1979) *Principles of Crop Improvement*. Longman, London and New York.

Simon, E.W. (1984) Early events in germination. In *Seed Physiology*, Vol. 2, *Germination and Reserve Mobilization*, ed. D.R. Murray, Academic Press Australia, 77–115.

Simpson, G. (1978) Metabolic regulation of dormancy in seeds—a case history in the wild oat (*Avena fatua*). In *Dormancy and Developmental Arrest. Experimental Analysis in Plants and Animals*, ed. M.E. Clutter, Academic Press, New York, 167–220.

Sondheimer, E., Tzou, D.S. and Galson, E.C. (1968) Abscisic acid levels and seed dormancy. *Plant Physiol.* **43**, 1443–1447.

Speer, H.L. and Tupper, D. (1975) The effect of lettuce seed extracts on lettuce seed germination. *Can. J. Bot.*, **53**, 593–599.

Steinbauer, G.P. (1937) Dormancy and germination of *Fraxinus* seeds. *Plant Physiol.* **12**, 813–824.

Steward, F.C. (1963) The control of growth in plant cells. *Sci. Amer.* **209**, 104–113.

Stokes, P. (1952). A physiological study of embryo development in *Heracleum sphondylium* L. I. Effect of temperature on embryo development. *Ann. Bot. N.S.* **16**, 441–447.

Stokes, P. (1953*a*) A physiological study of embryo development in *Heracleum sphondylium* L.

III. The effect of temperature on metabolism. *Ann. Bot. N.S.* **17**, 157–173.
Stokes, P. (1953*b*) The stimulation of growth by low temperature in embryos of *Heracleum sphondylium* L. *J. exp. Bot.* **4**, 222–234.
Stokes, P. (1965) Temperature and seed dormancy. In *Encyclopedia of Plant Physiology*, ed. W. Ruhland, Vol. **XV/2**, Springer, Berlin and Heidelberg, 746–803.
Sun, S.M., Mutschler, M.A., Bliss, F.A. and Hall, T.C. (1978) Protein synthesis and accumulation in bean cotyledons during growth. *Plant Physiol.* **61**, 918–923.
Thompson, K. and Grime, J.P. (1979) Seasonal variation in the seed banks of herbaceous species in ten contrasting habitats. *J. Ecol.* **67**, 893–921.
Thompson, K. and Grime, J.P. (1983) A comparative study of germination responses to diurnally-fluctuating temperatures. *J. appl. Ecol.* **20**, 141–156.
Thompson, P.A. and Brown, G.E. (1972) The Seed Unit at the Royal Botanic Gardens, Kew. *Kew Bull.* **26**, 445–456.
Thomson, J.R. (1979) *An Introduction to Seed Technology*. Leonard Hill [Blackie], Glasgow and London.
Totterdell, S. and Roberts, E.H. (1979) Effects of low temperatures on the loss of innate dormancy in the development of induced dormancy in seeds of *Rumex obtusifolius* L. and *Rumex crispus* L. *Plant Cell Environ.* **2**, 131–137.
Trewavas, A.J. (1982) Growth substance sensitivity: the limiting factor in plant development. *Physiol. Plant.* **55**, 60–72.
Vaughan, J.G. (1970) *The Structure and Utilization of Oil Seeds*. Chapman and Hall, London.
Vegis, A. (1964) Dormancy in higher plants. *Ann. Rev. Plant Physiol.* **15**, 185–224.
Villiers, T.A. (1972) Seed dormancy. In *Seed Biology*, ed. T.T. Kozlowski, Vol. 2., Academic Press, New York, 220–281.
Villiers, T.A. and Wareing, P.F. (1964) Dormancy in fruits of *Fraxinus excelsior* L. *J. exp. Bot.* **15**, 359–367.
Villiers, T.A. and Wareing, P.F. (1965) The growth-substance content of dormant fruits of *Fraxinus excelsior* L. *J. exp. Bot.* **16**, 533–544.
Warcup, J.H. (1980) Effect of heat treatment of forest soil on germination of buried seed. *Aust. J. Bot.* **28**, 567–571.
Wareing, P.F. (1965) Endogenous inhibitors in seed germination and dormancy. In *Encyclopedia of Plant Physiology*, ed. W. Ruhland, Vol. **XV/2**, Springer, Berlin and Heidelberg, 909–924.
Wareing, P.F. and Foda, H.A. (1957) Growth inhibitors and dormancy in *Xanthium* seed. *Physiol. Plant.* **10**, 266–280.
Webb, D.P. and Wareing, P.F. (1972*a*) Seed dormancy in *Acer pseudoplatanus* L.: the role of covering structures. *J. exp. Bot.* **23**, 813–829.
Webb, D.P. and Wareing, P.F. (1972*b*) Seed dormancy in *Acer*: Endogenous germination inhibitors and dormancy in *Acer pseudoplatanus* L. *Planta* (*Berlin*) **104**, 115–125.
Weier, T.E., Stocking, C.R. and Barbour, M.G. (1970) *Botany, An Introduction to Plant Biology*. 4th edn., John Wiley, New York.
Wesson, G. and Wareing, P.F. (1969) The role of light in the germination of naturally occurring populations of buried weed seeds. *J. exp. Bot.* **20**, 402–413.
Williams, P.M., Ross, J.D. and Bradbeer, J.W. (1973) Studies in seed dormancy. VII. The abscisic acid content of the seeds and fruits of *Corylus avellana* L. *Planta* (*Berlin*) **110**, 303–310.
Williams, P.M., Bradbeer, J.W., Gaskin, P. and MacMillan, J. (1974) Studies in seed dormancy, VIII. The identification and determination of gibberellins A_1 and A_9 in seeds of *Corylus avellana* L. *Planta* (*Berlin*) **117**, 101–108.
Wood, A. and Bradbeer, J.W. (1967) Studies in seed dormancy. II. The nucleic acid metabolism of the cotyledons of *Corylus avellana* L. seeds. *New Phytol.* **66**, 17–26.

Index